CATASTROPHIC TECHNOLOGY
IN COLD WAR POLITICAL THOUGHT

CATASTROPHIC TECHNOLOGY
IN COLD WAR POLITICAL THOUGHT

Caroline Ashcroft

EDINBURGH
University Press

Edinburgh University Press is one of the leading university presses in the UK. We publish academic books and journals in our selected subject areas across the humanities and social sciences, combining cutting-edge scholarship with high editorial and production values to produce academic works of lasting importance. For more information visit our website: edinburghuniversitypress.com

© Caroline Ashcroft, 2024

Edinburgh University Press Ltd
13 Infirmary Street
Edinburgh EH1 1LT

Typeset in 10/12.5 Adobe Sabon by
IDSUK (DataConnection) Ltd, and
printed and bound in Great Britain

A CIP record for this book is available from the British Library

ISBN 978 1 3995 3501 4 (hardback)
ISBN 978 1 3995 3503 8 (webready PDF)
ISBN 978 1 3995 3504 5 (epub)

The right of Caroline Ashcroft to be identified as the author of this work has been asserted in accordance with the Copyright, Designs and Patents Act 1988, and the Copyright and Related Rights Regulations 2003 (SI No. 2498).

CONTENTS

Acknowledgements	vi
Introduction: Catastrophic Technology in Modernity	1
1 Cold War Critics of Technology	23
2 Historical Narratives of Technological Development	63
3 Technologies of Destruction: The Shadow of the Bomb	96
4 Technologies of Production and the Rise of the Machine	118
5 The Veil of Technology: Media, Propaganda and Ideology	142
6 Technologies of the Body: Man as Raw Material	163
7 Technology and Worldliness: Nature and the Technological Artifice	186
Conclusion: The Lasting Influence of 'Catastrophic Technology'?	211
Bibliography	236
Index	248

ACKNOWLEDGEMENTS

Since I began working on this book in 2019, the world – and my world – has changed in ways I never imagined. The disarray caused by the Covid-19 outbreak posed its challenges for me, as for everyone, and along with personal losses, and on a brighter note, the birth of my two children, the past several years have been eventful. Among all this, however, it has been a privilege to have been given the time and space to write this book, and that is thanks first and foremost to the financial support provided by The Leverhulme Trust. I was fortunate enough to hold my Leverhulme Early Career Fellowship in the School of History at Queen Mary University of London, among a truly remarkable and talented group of colleagues. I am enormously grateful to all those in the school who have provided advice and support over the last several years, but particular thanks must go to Georgios Varouxakis, Waseem Yaqoob, Dan Todman, Andrew Mendelsohn and Quentin Skinner. To those who have taken the time to read and discuss all or part of the book, it is immeasurably improved thanks to your input. In particular, my thanks to those who took part in the workshop on this project, including Callum Barrell, Julia Costet, Robin Douglass, Hugo Drochon, Sean Fleming, Julia Nicholls, Max Skjönsberg and Daniel Wilson. My thanks must also go to Richard Whatmore, and the reviewers at *History of European Ideas*, as well as the reviewers for Edinburgh University Press, and to all those who have taken the time to offer comments on parts of the work that have been presented over the last few years, including at the ISIH conference in Edinburgh, the IHR seminar in London, the WPSA conference in Seattle and at the Institute of Political Science research seminar in Göttingen. As always, I am enormously grateful for Duncan Kelly's advice and support. Thanks to my friends, and finally to my family, Joby, Harry and Ada, and to my mum, whom we lost too soon.

INTRODUCTION: CATASTROPHIC TECHNOLOGY IN MODERNITY

An Idea of Technology

In the mid-twentieth century, a certain idea of technology emerged as a central category in the work of numerous political and social theorists, including many whose work was, or would go on to be, highly influential. This was a critical, even catastrophic, idea of technology, which was entangled with the apocalyptic fears engendered by two all-consuming world wars and the looming nuclear threat. A view of modern technology as being inherently dangerous was subscribed to by thinkers as diverse as the German philosopher Martin Heidegger and his students, Hannah Arendt, Hans Jonas and Günther Anders, the Frankfurt School theorists Theodor Adorno, Max Horkheimer and Herbert Marcuse (also a student of Heidegger's), French theologian and sociologist Jacques Ellul, and the American social and political thinker Lewis Mumford. The work of many other thinkers in this era reflects aspects of this idea but here this group will be the focus of attention, because their thinking represents the most emphatic, consistent and influential elucidation of the concept of technology as catastrophe. Modern technology, they believed, was a profoundly concerning form of technology; modernity had produced a particular kind of technological practice that threatened the very societies that it had emerged from. The catastrophic nature of technology meant that it was inexorably destructive of the political and social structures of human life, and also threatened the existence of human life itself through the introduction of cataclysmic risks such as nuclear war, global pollution or genetic manipulation.

Technology in the contemporary world cannot be understood as simply the aggregate of technological artefacts or processes, these thinkers argue, but possesses a particular character: 'technology' describes a type of process that can then be seen in those individual artefacts of technology. 'The essence of technology', as Heidegger famously wrote, 'is by no means anything technological.'[1] This modern practice of technology is inherently totalising and dominating, these thinkers argued, and destructive of what is most valuable about human life and society: difference, plurality and critique, and the freedom that emerges from and relies upon these political possibilities. As such it poses a definitive threat to politics. 'Technology is stronger than politics,' wrote Hans Jonas. 'It has become what Napoleon said politics was: destiny.'[2] This book asks the question of what technology means for these thinkers in political terms: what they understood the political impact of technology in modernity to be. As such, it explores, not technology per se, but a particular idea of technology and its implications in the work of the twentieth-century political theorists who subscribed to this idea of what I describe as 'catastrophic technology'.

Technology is a difficult concept to categorise because it lies between the idea and the material object and, equally, connects these two forms. Technology does not determine action. That, for instance, 'guns [alone] don't kill people' is objectively true. But technology clearly influences subjects, biases their action in certain ways, and enables certain actions, thus shaping the possibilities of action in politics and social interactions and, therefore, our understanding of the world. This combination of materiality and ideological influence is what makes technology such an interesting and complex subject of study within the field of political theory. In the early 1990s, three historians of technology wrote that, 'today it is obvious to virtually everyone, as it was not as recently ago as thirty years, that there is a "problem of technology"'. Indeed, they continued, it is 'poised to become our central concern . . . [and] if the most pervasive and profound characteristic of the modern age is the ever-expanding conquest of nature, then it must necessarily follow: the problem of technology is the problem of our time.'[3] Two points might be made here. First, that in the era of mass awareness of man-made (technologically produced) climate change, the challenges of the digital era, and the total ubiquity of technology in the everyday life of the average person (particularly, but by no means exclusively, in Western societies), it does indeed seem that the problem of technology has never been more intensely felt. The second is that, even if this seems to be the case, there were certainly those – our twentieth-century critics of technology – who felt that the problem of technology was the central problem of *their* time. In this context, it seems relevant to listen to what they have to say.

The narrative offered by these critics of technology emphasises the darkest side of technology's influence on politics, arguably to the exclusion of any *positive* influences it may have. Nonetheless, though the analysis is extreme in its scepticism towards technology, this approach also effectively underlines many characteristics of modern technology, connections between technology and politics, and the intractable problems and challenges entailed by contemporary technology, in ways that are illuminating to think through, even if *themselves* problematic or incomplete. Modern technology, seen as 'catastrophic technology' in this body of work, represents not only a form of ideology, as well as a mode of action, but also, more specifically, a kind of organisational pattern that merges with human activity across every sphere of human existence, that becomes continuous and contiguous with human action. The depiction of technology offered here is not that of a straightforwardly anthropomorphised 'machine' technology where technological beings come to dominate humanity, even if there are overtones of this at times. It is not simply that the machines take over. Rather, the critics of technology offer up a complex understanding of how technology becomes increasingly intertwined with life in all its forms, and in which the relationship becomes mutually interdependent. These thinkers believe that this interaction is fundamentally problematic, in ways that this book will set out, but one need not necessarily agree with their claims in extremis to find their considerations on the nature of the relationship between technology and politics in modernity valuable. They offer ways to think through how technology influences politics and social interaction in the widest sense. It is also the case that many of the interactions our thinkers outline no longer operate in the same way, or take the same form in twenty-first century technology. Many of their predictions about the catastrophic qualities of technology have not – yet, at least – come to pass. On the other hand, many of the fears they express sound familiar in the context of contemporary warnings of environmental catastrophe. In environmental politics we find echoes of many of their fears and claims, a correspondence which is not coincidence, as this book will show, but reflects the intellectual influence of the technological catastrophists upon environmentalism in its nascent stages.

This book attempts to outline this particular historically situated notion of technology, and the critique of politics connected to it, through exploring the overlapping and intersecting claims and beliefs of some of the more important thinkers who subscribed to this idea. The centrality of technology in the work of theorists such as Heidegger, Ellul, Jonas, Adorno, Horkheimer, Marcuse and so forth, is well documented, and in certain ways their ideas have been considered in relation to one another, particularly those of the Frankfurt School on the one hand, and those of Heidegger and his students on the other (although, complicating matters, Heidegger's influence on some members of the Frankfurt School was itself significant). Don Ihde describes the 'emergent theme of much mid-twentieth-century philosophy of technology in both Europe and America' as *'autonomous technology*, that is, a runaway technology that exceeds, "Frankenstein-like," its inventor's controls', characterising it primarily as a response to, or development of, Heidegger's philosophy of technology by his former students.[4] This reflects that, in the history of the philosophy of technology, there has been a tendency to focus on the contributions of a few particularly notable philosophers of technology in this period: most importantly Heidegger, Ellul and, to a lesser extent, Marcuse. There is no doubt that these figures are of seminal importance in the history of the philosophy of technology, and that Heidegger's influence in particular is profoundly important – 'the thinker who has most deeply pondered our technological destiny'.[5] But the questions that they grapple with, and the technological critiques they offer, were part of a much more widespread engagement that cannot be simply traced back to these few thinkers. As such, the focus of this book is the wider historical influence of this idea of technology as a significant narrative or discourse in the Cold War era, and it attempts to draw out the character of this technological idea as it emerges and develops in the mid- and late twentieth century. The figures that form the focus of this research are well known as critics of modern technology. Yet while the critique of technology is of major importance to many pre-eminent thinkers of the twentieth century, offering some innovative perspectives on Cold War politics and ideology, as a substantive theme in the political thought of the period, its importance has been underestimated in existing intellectual histories.

The Political Character of the Technology Critique

These theorists' work on technology is essentially political; to think about technology in their work is to understand their political fears in a more nuanced way. This critique of technology is, to a significant degree, a critique of liberalism, understood as the politics of the contemporary Western world and the primary ideological political orientation of the post-war West: the political regime that claims to prioritise the liberties of the individual, within a constitutional and pluralistic order. Liberalism itself can be defined in many different ways, but for the critics of technology it is generally understood in the negative rather than possessing a clear positive content: they forefront what it is *not*, conceptualising it in terms of the problems it entails for politics. They attack, especially, idealised perceptions of liberalism. One claim that can be identified in the work of many of the critics is that liberalism has been superseded by technology, and that whatever benefits liberal politics *had* provided are now in the past. 'As a personal and social philosophy, liberalism has been dissolving before our eyes only during the past decade,' wrote Mumford in 1940.[6] 'The more technical progress advances, the more constricted will be the role of liberalism,' argued Ellul. 'Liberalism is only conceivable if technical progress is choked off.'[7] Adorno and Horkheimer distinguish the 'phase of mass culture' from the 'late liberal' stage that *preceded* it, writing that liberal theory serves now only as an 'apologia for existing circumstances'.[8]

4

In the context of Cold War America, at a moment when the moral and ideological power of liberalism was arguably at its height, this perspective is incongruous. It draws in part on the pervasive and varied Weimar-era critiques of liberalism that formed the background to the theorists' earlier lives, and indeed the abject failure of liberalism in Germany in 1933. But the claims about technology presented here have a particular relevance in the political context of the Cold War, and can be framed as a response to Cold War politics. That is, in the context of an apparently bipolar political world, these theorists claim that there is a more fundamental influence on global politics than the superpower geopolitics that is taking place between the Soviet Union and the United States: technologisation, which they explicitly consider to be essentially the same in both the communist and democratic worlds, is equally politically concerning, albeit taking different paths of development. Later works, Arendt's *The Human Condition* as well as her critical commentary on modern culture or space travel, for example, or Ellul's critique of political propaganda, also argue that the Cold War has extended the use of technology in new and uniquely dangerous ways.

While this is therefore a fundamentally anti-totalitarian critique, it is distinct from the anti-totalitarianisms of, for example, liberal philosophies that saw communism and Marxism as the major threat, such as that of Karl Popper; or for those who, like Friedrich Hayek and Milton Friedman, made the case for the liberalisation of markets on the basis of guaranteeing liberty; or the French anti-totalitarianism of the 1970s, which saw large sections of the French left reject revolutionary Marxism, in a party-oriented critique of communism. The technology critique in fact minimises the significance of such distinctions in arguing that the relevant problem is that of technology, in the West just as much as the East.

Duncan Bell has highlighted the way in which contemporary ideas of liberalism developed in the mid-twentieth century in response to growing fears about the threat totalitarianism posed, and Samuel Moyn has also recently written on how this 'Cold War liberalism' emerged in the middle of the century in a novel and particular form, distinct from earlier ideas of liberalism, but claiming the label.[9] The critics of technology, taking the totalitarian threat just as seriously, identify contemporary liberalism as a fundamentally outdated or superseded form of political action that acts primarily to *mask* the seriousness of the threats that technological modernity poses to politics. But liberalism can also be read here, explicitly against the 'anti-totalitarian liberalism' that was so significant in this period, to have a more active influence in stimulating technological development, and its totalitarian tendencies. A key tenet of liberalism, as these theorists understand it, is the idea that history is progressive and that liberal societies are essentially and necessarily progressive. For the critics of technology, 'technology' is almost synonymous with 'progress', and thus we can draw the connections between technology, progress and liberalism. The same point, it should be said, can be made of the self-perceived progressiveness of communist politics in the Soviet Union – in both cases, an accelerating progress is seen to animate political action, and thus technology and politics are in lockstep. As such, technology cannot be considered politically neutral, suggest these theorists, but rather technology is thoroughly political *in* its progressiveness. The technology critique thus offers a different entry point and perspective for thinking about Cold War politics than other more familiar approaches. Ideological positions, political strategies or economic approaches, and even the authority of the state, appear as peripheral to or derivative of technological processes of change and development.

Whether or not one agrees that modern technology is catastrophic in the ways these theorists claim, their analysis highlights the way in which technology in modernity is an intrinsically political concept. Their arguments highlight the impossibility of treating technology as something neutral; it shows the profoundly political nature of the concept of technology and the impossibility of extracting or excluding technology from political narratives or political critiques of the modern era. 'To see technology as "just another problem," as by no means fundamental, is to miss its obvious character as a horizon within which every other problem comes to light,' writes Jerry Weinberger. 'The possibilities of technology constitute not just any modern world view; they refract our experience of any possible world, whether it be past, present, or in the future. It is impossible to conceive of modernity without reflecting on technology.'[10] The experiences of modernity, Langdon Winner observes, show us that 'technologies are not merely aids to human activities, but also powerful forces acting to reshape that activity and its meaning'.[11] This can be seen, he explains, in

> instances in which the invention, design, or arrangement of a specific technical device or system becomes a way of settling an issue in the affairs of a particular community . . . [or in the case of] 'inherently political technologies,' man-made systems that appear to require or to be strongly compatible with particular kinds of political relationships.[12]

It is dangerous, as George Grant has pointed out, to 'prejudge' the question of what technology is, by characterising it as nihilistic.[13] The work of the critics of technology should not simply be taken at face value for this reason (among others). But while their idea of technology as essentially catastrophic shapes their analysis of technology's role in the world, *all* analyses of technology, including those which portray it as a more neutral or positive force, address a concept that is now firmly embedded in modern politics at every level and thus is inherently political.

In Leo Marx's history of technology as a 'hazardous concept', he suggests that the very reason for the triumph of the concept of technology is its 'vague, intangible, indeterminate character – the fact that it does *not* refer to anything as specific as a tool or machine'.[14] It is hazardous, he writes, because 'the fact that we cannot say what the word means' limits our capacity for analytical thinking.[15] And the attribution of agency to technology, the suggestion that technology is the 'primary force shaping the postmodern world', is a factor in 'our growing reliance on instrumental standards of judgment, and our corresponding neglect of moral and political standards'.[16] The critics of technology undoubtedly fall into the trap of attributing (at least a degree of) agency to technology. But so, Marx suggests, does almost everybody, whether critically or favourably. To understand the political character and implications of the technology critique is an essential aspect of assessing the critics' ideas of technology. These ideas are integral to the politics and philosophy of these intellectuals, and many of the elements of this critique of technology have also been translated into the political imagination of the contemporary world in various ways, not least through the environmentalist critique of technology which, as this book will show, not only is in alignment with, but also draws upon these ideas.

The work of philosophers of technology Langdon Winner and Andrew Feenberg have engaged with precisely the kind of technological critique outlined here. The present work seeks to develop the kind of themes these theorists are engaged with, deepening our historical understanding of the position of the critique of technology in this period.[17] Andrew Feenberg's work seeks out a path between the technology critiques of Heidegger and that of Marcuse in his development of a contemporary critical theory of technology.[18] In this project he highlights how 'after World War II, the humanities and social sciences were swept by a wave of technological determinism. If technology was not praised for modernising us, it was blamed for the crisis of our culture.'[19] Thinkers including 'Weber, Lukács, Heidegger, Foucault, Mumford, Horkheimer, Adorno, Habermas, and Marcuse created a . . . modernity critique in which science and technology are grasped not as a specialized activity or a productive force but as a cultural phenomenon,' he writes.[20]

Winner has also written extensively on the idea of technological critique, which he describes as 'autonomous technology', or 'technics-out-of-control'. Drawing on an array of twentieth-century thinkers, including (but not limited to) Mumford, Marcuse and, most importantly for him, Ellul, Winner writes that in their works 'one finds a roughly shared notion of society and politics, a common set of observations, assumptions, modes of thinking and sense of the whole which, I believe, unites them as an identifiable tradition'.[21] The tradition that Winner identifies is precisely that treated in the present study, and the overarching characteristics are as presented here: this idea presents modern technology as uncontrollable, subject to exponential growth, and productive of an array of social and political threats. However, I argue that the critique of technology, in a genuinely coherent form, should properly be understood as a Cold War phenomenon, albeit one which draws upon pre–Cold War debates in its depiction of technology and its positioning of technology as a specifically political phenomenon. It is notable that Winner's key protagonists are all writing in the Cold War era, and this political context is important to this technological discourse, particularly to the more extreme depictions of modern technology's catastrophic or apocalyptic qualities. Such 'apocalyptic potential', wrote Jonas, includes 'its ability to endanger the very existence of the human species, or to spoil its genetic integrity, or arbitrarily to alter it, even to destroy the conditions of higher life on earth'.[22] This work thus develops our understanding of the tradition that Winner and Feenberg identify and draw upon, by historicising it, and by exploring in greater depth the historical specifics of the critique of technology in its various dimensions. That is, this work seeks to analyse the way that this group of Cold War–era political thinkers and philosophers critique different forms of technology in the world around them, and the challenges this poses to politics and political ideals. It examines not just the overarching 'system' of technology these theorists describe, but also how they believe this system acts upon various aspects of politics – the capacity of technologised politics to offer stability and permanence, the degree to which it permits individual autonomy and the pursuit of meaningful living, the relationship between the public and private spheres, the existence of a world of common meaning and the potential for political critique, the relationship between nature and politics – through specific means and forms of technologies.

Recovering an Idea of Technology

It is valuable to recover this idea of technology or, more specifically, to recognise its importance in the political and social theorising of the Cold War era because the critique of technology offered by these thinkers is nothing less than a means of reimagining the modern world. This takes various forms. First, a focus on technology in the work of these theorists helps us to better understand their criticisms of technology as a novel response to Cold War geopolitics and a contemporary reimagining of the Cold War world. The development and utilisation of technology is a global trend, they argue, with fundamentally the same effects in different parts of the world, whether East or West, communist or liberal democratic. Because the influence of technology on politics has universal characteristics, it is also a globalising force, and one which therefore unites even the Western world with the communist world. Jerry Weinberger has written of how, for liberals, the dangers posed by technology, as far as those dangers exist, are 'posed somehow from without . . . conceived as the unintended consequences or "externalities" of the well-meant and legitimate pursuit of happiness and security'.[23] This work explores a group of thinkers who sought to reframe technology as intrinsically related to *all* modern forms of politics, liberal or authoritarian, democratic or communist.

The critique of technology reveals a particular kind of world, one that has subsumed all other possible political, social or cultural worlds in modernity, that which is produced or enabled by technology and the technological process. The technological world is in some sense a 'neutral' society, that is, absent of critique, dissent or plurality – the qualities which enable politics or political freedom for these theorists. It is, equally, a total or totalising society, in part because of the absence of difference and critique, but also because technology is, these thinkers argue, oriented towards the domination of nature, and thus the domination of humankind. 'When technique enters into every area of life, including the human, it ceases to be external to man and becomes his very substance ... it progressively absorbs him,' writes Ellul.[24] Modern technology negatively unifies the world. This conceptualisation of the technological world, from a wide spectrum of political and social theorists with radically different political commitments, highlights that, despite their differences, there are commonalities here that are important to how they conceive of politics or society. And in the context of twentieth-century thinking on the global environment in which we live, this perspective foreshadows later notions of the Anthropocene. Benjamin Lazier connects the widespread acceptance of terms such as 'globalization', 'global environment', 'global economy' and 'global humanity' with the *Earthrise* images of the late 1960s and changes in the 'Western pictorial imagination' that followed those images.[25] From the 1960s and 1970s, a concept of the Anthropocene begins to emerge which sees the world as a unified whole, an inherently political concept, as Duncan Kelly has highlighted.[26] However, the development of such conceptions of a technologically unified globe is clearly also identifiable in the work of thinkers such as those explored here, including in work which precedes the '*Earthrise* era' and even more so the Anthropocene concept. These technological sceptics contend that technology is changing how we perceive, interact with and create our world, and the relationships between humans and nature. Technology, they argue, increasingly mediates, and sometimes even replaces, human relationships with the world: with political authority, with one another, with one's environment. Their work thus represents an early but sophisticated analysis of the way that technology is – intentionally, but more often unintentionally – changing the fabric of the world we live in, and how we might understand and cope with that.

By the 1960s and 1970s, similar concerns became much more widely adopted, as the work of early environmental thinkers like Barry Commoner and Rachel Carson, and burgeoning green movements, began to gain traction.[27] The critique of technology, and particularly the conceptualisation of technological worldliness that comes from that critique, offers another perspective on the impact of human action – particularly our technology – on the environment. This discourse influences and is in turn influenced by more overtly environmental works that are written during the later Cold War. The technology critique thus intersects with the early popularisation of ecological thinking and environmental movements and politics. In the present context of a continued and indeed ever increasing interest in the relationship between technology, world and the political ramifications of such change, this is therefore a discourse which is of particular interest to the contemporary world – even if the context and immediate concerns, both political and technological, differ from our own. The question of how technology both creates and equally damages the world in which we live is at the heart of these theorists' analyses.

This project is then, not a history of technology, but a history of technology as a political concept during the Cold War: as variable, contested and dependent upon positions of ideology, belief, and perception as any complex political concept. Technology was used by these thinkers not just to represent a key threat to political norms or ideals – freedom, democracy, the rights of the individual – but the concept was also therefore used by them as a rhetorical weapon. The focus here is on technology as a political concept and thus the technological discourse is read as polemical discourse; an attempt to intervene in order to revise and augment existing meanings of technology, particularly against what the proponents of this technological scepticism saw as dominating liberal-progressive representations of technology. It explores the way in which a set of thinkers are engaged in critically developing and re-imagining 'technology' as a political concept or issue, not only in the abstract, but through their analyses of particular technologies or forms of technology and their political impact, as this book will show. In their work, 'technology' or 'the technological' acquires a particular political meaning as a unitary and coherent concept, one which is explicitly understood to be a modern formulation; particular political and social meanings are embedded within this terminology. This is prefigured in some pre–Second World War thinkers in Germany such as Weber, Jünger and Spengler, and even Heidegger's pre-war work to some extent, but it becomes much more highly developed after the war, not least because of the international influence of the German intellectual emigres who picked up, developed, or responded to some of the themes of some of these earlier writers. Specifically, the Cold War thinkers are developing technology as a political concept in opposition to a perceived notion of technology as a positive political concept (or technology as 'progress').

Although primarily a historical study, it is also the case that the questions and concerns that these theorists raise are still of contemporary relevance. It is inevitable that when thinking about relatively recent texts, which address questions which remain so pertinent to today's politics and society, parallels might be drawn. This study does not answer these questions, nor does it assert that the fundamental issues are the same. But there are continuities (as well as discontinuities), and the work, I hope, will provide a new historical perspective on the theme of technology in politics. One need not agree, or disagree, with the ideas expressed by these thinkers, in order that their work might help one to think through related issues. It is not the primary focus of the work to provide an analysis of contemporary technology and politics, but it ought to be recognised that the drawing of parallels is an inevitable quality of such a study, and one that can be useful, as long as the unique qualities of the period and discourse itself are also considered.

'Technology': Some Conceptual Clarification

The term 'technology', as used in this book, is a shorthand for an interconnected complex of ideas found in the work of these authors, and which incorporates other terminology as well, and a multitude of (actual) technologies. Different terms or words are used, across several languages, with somewhat different intonations: technology, technics, technique, *Technologie*, *Technik*, technicity, technological society and many more variations. Much could be said about the etymology. In relation to French terminology, Michael Behrent highlights the distinction between the *la technologie* and *la technique*, the latter being the more expansive term. Delanty and Harris observe that those within the critical theory tradition, as well as Weber and Heidegger, adopted *Technik*, the dominant nineteenth-century term which referred to instrumental technology, rather than the older *Technologia* which had applied to the technologies of the arts and had closer links with the Greek *technê*. Eric Schatzberg argues that technology in its present incarnation became part of the English language as a translation of the German *Technik* around the turn of the twentieth century.[28]

More tangibly, 'technology' refers also to different forms: military technologies, bioengineering and medical technologies, media, communications and culturally influential technologies, technologies of production and automation, among others. What is significant, and what permits the general use of the term 'technology' in this study to tie together the very different practices and their impacts, is that the group of intellectuals studied here all consider these technologies to possess similar characteristics, or operate through similar processes – that is, in a similarly expansive, theoretical and generalising way. This is technology understood as a mode of action, a type of process which is linked to material technologies. Thus, while this work will explore a range of particular technologies and their social and political impact, this study also seeks to thereby illuminate an overarching set of interlinking theories of 'technology' and its relation to politics. Similarly, just as the theories of technology and their place within philosophical understandings of politics and society vary between thinkers, so too does the precise terminology they use. But, on the whole, we can identify substantial overlap between these writers' ideas of technology, and it is this convergence that this work is concerned to illuminate – without excluding clear variations between their perspectives.

Some preliminary points of orientation should be made. First, the concept of 'technology' referred to here is modern. Although technology has been a component of human life since its prehistory, and a fundamental aspect of the way that individuals relate to the world and each other, it is not used by these thinkers as a timeless notion. They recognise the history of technology (as tools), but the technology they are interested in and concerned by is the technology of the modern world, and the technological processes that emerged primarily, but not exclusively, in the nineteenth and twentieth centuries. The history of technology that they are interested in precedes this but, for the most part, only by a few hundred years, back to the start of the early modern period. That means that, when they refer to technology, there is a tendency to be somewhat blind to the continuities of tool usage and technological utility, and to focus on its novel aspects. This shapes their idea and critique of technology in important ways.

Second, the character of technology as outlined in this conceptualisation of technology is fixed or determined only to the extent (and this is significant) that modern society continues to develop and perpetuate itself according to its present trajectory. That is, the problem of technology, for these theorists, is not inherent in technology itself, but specifically in the interaction of technology with modern society. Technology as it has emerged and continues to evolve within modernity has taken a certain shape not as a result of any inner 'essence' (although they certainly identify important trends and tendencies) but because of the shape of modernity itself. It is the confluence of modernity and technology that is so politically worrying for these thinkers, and it is this that they emphasise in their analyses. Technology is socially determined, but so too does the direction of travel of technology in such societies lead to certain almost inevitable consequences (they argue). Thus, the qualities of technology are as malleable or immovable as modernity itself. The thinkers themselves universally proclaim the possibility and necessity – though not the likelihood – of change, for instance Heidegger, who seeks out the overcoming of the technological 'enframing' of reality; Arendt, who makes the case for a kind of political revolution that looks back to classical traditions of freedom; or Marcuse, whose political activism spells out the possibility and necessity of the dialectical transformation of modern culture.

The final point relates to science. Technology here is understood to be distinct from, although closely related to, both science and rationality or the process of rationalisation. Writing in the 1980s, Stanley Aronowitz wrote that most 'investigators of science remain tied to the concept of science as a distinct knowledge sphere and have barely touched its relation to technology'. The great exception to this, he argues, were the critical theorists, who instead made the relationship foundational to their philosophy.[29] While this description of critical theory's idea of technology is accurate, the idea that the critical theorists

INTRODUCTION

were alone in making this claim is incorrect, as this book will show. All the critics of technology, not just the critical theorists, emphasised the importance of technology above science in the modern era, and its dominant influence on society. The importance of technology lies in its material aspect. Unlike science which is firstly or primarily abstract or ideal (as rationality), technology is worldly and this is highly significant for the political implications of the technology critique, specifically through its exploration of the political effects of technology not just in the abstract but through specific types of technology.

KEY INFLUENCES ON THE TECHNOLOGY CRITIQUE

Technology is a political concept and, like all political concepts, must be understood to be contextually specific. Thus, this notion of technology emerges as a shared narrative against the backdrop of the Cold War, but was also influenced in important ways by the interwar years and the events of the Second World War. There is a particularly German influence on the idea of technology presented here: the body of work in which this concept of technology can be found is dominated particularly by German-Jewish thinkers, and, although a Cold War–era concept, it is heavily influenced by – or we might consider it a response to – earlier German thinking on the interaction between technology and society, and indeed the path of German politics itself.[30] The significance of German thinkers might speculatively be attributed to several influences. The dominance of Germany in European intellectual life in the early twentieth century is likely one contributing factor. The importance of technology in German thought itself might be to do with the relatively late industrialisation of Germany and its unification, nationalisation and militarisation in the late nineteenth and early twentieth centuries. The discourse on technology that began in the early nineteenth century in Germany continued to influence many thinkers including those studied here. It is by no means, however, an exclusively German idea, but can be found in the work of French and North American authors, and the exodus of German intellectuals to the rest of Europe and to America means that the national context of the emergence of this idea is anything but straightforward. The historically unique position of the German theorists themselves, many of whom were exiled as Jews (some, but not all, later returning to Germany) and many of whom became important, even transformative, figures in the academic spheres of their adopted homelands, particularly, of course, the United States, means their work transcends national borders. The idea of technology that appears is, therefore, an idea that, if not truly *global* (because it remains, largely, Western-centric), certainly travels beyond Germany through several paths, and is also rooted in a critical response to American politics and technology that appears in the work of these thinkers.

In the philosophy of technology in the twentieth century, Heidegger has been a dominant figure. Given this fact, any history of the technology–politics

13

relationship must necessarily take account of Heidegger's work and his profound influence on other thinkers. In this study, however, he is not the dominant figure by any means, but one among many, and, in certain respects, of less interest to political questions. Heidegger is primarily concerned to find the origins of philosophy, the basis of truth or understanding, rather than being concerned with the particulars or specifics of what technological change might mean in political practice (although certain technological examples were nonetheless influential as he developed his critique of technology). This work seeks to focus much more on the particulars of Cold War technology and the ideas and critiques of such technology – with the observation that theoretical, ideological and philosophical ideas cannot be divorced from how these technologies and their political importance or influence were understood. It is also primarily a study of political ideas rather than philosophy in a broader sense. In short, this account offers a more materially contextualised account of the ideas of the relationship between politics and technology. In fact, I do not take any particular thinker to be dominant in this story, but rather through the presentation and analysis of multiple and overlapping contexts – intellectual, ideological, political and biographical – seek to present an intellectual interaction as it emerges between and as a result of these different contexts, 'eavesdropping' on the conversations of the past, as John Burrow once described the practice of intellectual history.[31]

This collection of intellectuals who identify technology as being a major threat to modernity is evidently comprised of individuals who occupy in many respects radically different political and philosophical positions from one another. They can be considered as a coherent grouping (for the purposes of this work) in the sense that their work appears to draw on a notion of 'technology' with substantial commonalities; substantial enough, this study contends, that we might understand this particular idea of technology as being a significant concept in political theory at this time, as a conceptual context for their work, and thus as a narrative that these thinkers spoke to and sought to shape through their work. Sometimes, these thinkers were literally in dialogue with one another, citing, critiquing or responding to others' ideas; for others, there is no clear line of influence or relationship. Just as significant, therefore, were the shared concerns or common starting points that can be identified among this group that fuelled this shared conceptualisation of technology. Despite their ideological diversity, spanning the political spectrum from far left to the furthest right, they share the intellectual heritage of the early twentieth-century German, and especially Weimar, academic milieu, in all its variety and complexity. Many of these individuals also share the personal experience of exile from Nazi Germany and immersion in American society, although this again, of course, also took on different forms and meanings in their understanding of political modernity. The group is notable, in other words, for the fact that, despite the intellectual and sometimes personal differences or even antipathies between them, they share in the adherence to a substantially similar and interconnected narrative of technology's meaning and role in modern politics.

Nonetheless, certain political – or perhaps, ideological – commonalities between these individuals can be found, however unlikely that might seem at first sight. Specifically, the political contention that connects these thinkers is their mutual critique of liberalism: specifically, in relation to technology, a critique of the presumed 'neutrality' of technology by liberal philosophy and the idea of progressive politics that furthers the technological ideal while at the same time hiding the increasing dominance of technological threat or risk. The critiques come from very different political perspectives, from the Frankfurt School's neo-Marxism, Arendt's republicanism, Ellul's anarchism or Heidegger's far-right, erstwhile fascist, nationalism. These might even seem too radically different to be said to pull together in any meaningful sense; they certainly describe liberalism in distinctive ways, even if their critiques overlap on the influence of technology. Yet all of the theorists in the wider group this work explores believe that liberalism – whatever the original intention or virtues attached to it were – has been, likely irredeemably, perverted by the course of modern society and its political and economic structures, a modernity shaped as much by technology as by anything else.

The development of complex narratives of the influence of modern technologies on politics can be traced back at least as far as the Industrial Revolution: from critical, wholesale rejections of technology to utopian visions of its possibilities. Yet, as highlighted by Thomas Hughes, 'the word "technology" came into common use during the twentieth century, especially after World War II'.[32] The idea of technology this book concentrates on is fundamentally of the Cold War era in certain key respects: the publication period of many of its key texts, the technologies it refers to, and the political context it speaks to, but it also draws on earlier theorisations of technology, and is considered by these theorists to be relevant to the modern era more broadly. That is, the idea of technology outlined here is a critical commentary on the immediate global political Cold War context these thinkers find themselves in, as well as the politics of 'modernity'. Their histories of modern technology date back to early modernity in many cases, and thus interact with their histories and critiques of the development of concepts such as the modern sovereign state, liberalism, democracy and science. Arendt states this succinctly in her statement that the atomic bomb, 'though it represented something absolutely new in the history of science', was 'nothing more than a culminating point, achieved, so to speak, by one short jump or short circuit, toward which events in any case had been moving at an ever accelerating pace'.[33] Technology's influence is interwoven here with the development of 'progress', a term which is seen in these texts as almost synonymous with technology, and which connects the idea of technology to these thinkers' critiques of liberalism.

We can trace the intellectual roots of this Cold War reading of technology back to the 1920s and 1930s, with the rise of fascist and communist totalitarianisms. With the Cold War, these theorists' fears of totalitarian society move beyond specific national threats, and through technology totalitarianism is seen to become a universal threat posed by modernity itself. 'The world in which we live today and which surrounds us, is a technological one,' insists Günther Anders; we live in a technological condition.[34] Modernity, insofar as it is technological, is inherently totalitarian. 'In the face of the totalitarian features of this society, the traditional notion of the "neutrality" of technology can no longer be maintained,' Marcuse wrote in *One-Dimensional Man*.[35] Technology makes the threat of totalitarianism, or, in even more general terms, the threat of apocalypse, a threat posed by modernity itself, not by particular regime forms or ideologies or individuals. Modernity possesses a systemic tendency to self-destruction. The narrative about the connection of technology and politics explored here is one of distinct and increasing pessimism, in contrast to pre–Cold War thought, which is much more diverse in its perspectives, or in contrast to what the critics of technology portray as the wider contemporary context of social optimism about technological progress, certainly in the earlier decades of the Cold War. By the late 1960s, 1970s and 1980s, the mood expressed by the technological pessimists has grown more widespread, and this is an important part of this history, as concerns over the technological influence on the environment become more widespread.

It is certainly not the case that the idea of technology and its implications, with which this book is concerned, was a *comprehensively* agreed upon notion. While their ideas overlap in important respects, each theorist offers their own portrayal of technology, drawing on different instances and aspects of technology in modernity. Even less so was it the case that this idea, to the extent that it did form a coherent critique of technology, was more broadly undisputed in this period – far from it. Numerous technological optimists offer very different narratives: the anti-liberal and anti-democratic technologism of Ernst Jünger and Oswald Spengler in 1920s and 1930s Germany; John Dewey's vision of technologically enhanced democracy in early twentieth-century America; Stewart Brand's 1960s countercultural appeal to an individualised, empowering technology; Buckminster Fuller's belief in the almost limitless technical possibilities for 'spaceship Earth'; Harold Wilson's vision of a Britain transformed by the 'white heat' of technological revolution; Shulamith Firestone's 1970s feminist fantasy of a future where reproductive technologies might bring about the end of sex distinction itself, and its associated inequalities; or those inspired by the new science of cybernetics who rejected technological anxieties by virtue of the claim 'that biological organisms and modern machines were two instances of the same phenomenon'.[36] These are just a few among the very many who saw great potential in the future of technology for human politics and society.[37] In fact, it was the opinion and fear of most of the technology critics that technology was generally considered in positive terms in modernity, or at the very least as a neutral force. As John McDermott wrote in 1981, the belief of neutrality in politics or what he describes as a '*Laissez innover*' approach assumes that 'technology is a self-correcting system . . . [and that] "negative externalities" will and should be corrected by technological means'.[38] The claim that technology is neutral, in other words, is a strong claim towards the development of more, and unrestricted, technological development.

A Different Globalism

Although the claims of the critics of technology are informed by earlier debates around technology, and their idea of technology is situated in the broad historical context of modernity, the political critique embodied in their concept of technology is a response to the specific period in which these texts are written. The discussion of technology and politics offered here is a sustained critique of modern politics via the particular critique of technology; a reaction to the modern political world, and to its dominant political philosophies or ideologies – be that liberal-democratic or communist. These thinkers implicitly and sometimes explicitly oppose thinking in terms of bipolarity as the primary method of conceptualising Cold War politics. A number of these philosophers simply reject the Cold War dichotomy of 'us and them' or 'East and West', in favour of a world view that instead identifies fundamental affinities between Western 'progressive' or liberal societies and Soviet communist societies. Marcuse, for instance, argued that there is 'an essential link between the two conflicting [Soviet and Western] systems . . . the technical-economic basis common to both systems, i.e. mechanized (and increasingly mechanized) industry as the mainspring of social organisation in all spheres of life'.[39] *Both* of these regime forms are worryingly prone to totalitarian or proto-totalitarian tendencies. The political poles of existence, for these thinkers (in the context of technology), are not left and right, or liberal and communist, but rather totalitarian and non-totalitarian. The idea of the moral or political supremacy of the West is also questioned by these thinkers, who clearly came from a Western tradition of political thought, and they see no necessary reason for the future success or dominance of liberal democracy. They were opposed to both the flawed ideologies of liberalism and Soviet communism, and a significant aspect of their criticism of both regimes is the use and understanding of technology in these political communities; in both, technology is seen as a repressive modern force enabling and driving both liberal and communist politics, in a trend that tends in both situations towards political catastrophe or emergency.

Technological Catastrophe?

These intellectuals feared the impact of technology on politics, and their collective work on this theme represents a substantial critique of what they considered to be a likely and perhaps inevitable technologically driven catastrophe. More specifically, it is not technology itself or technology as such that is seen to determine humanity's future, but rather the way in which technology appears and is operationalised within contemporary society. Thus, the technological catastrophe need not take the form of nuclear apocalypse, or similar, although that is one potential future. Catastrophe is often framed as a violent and destructive, dramatic 'overturning and overthrowing'.[40] The sudden rise to power of a totalitarian regime, or the immediate impact of a nuclear war, would fit this model. But the catastrophe of technology can also be conceived in other ways that are also comprehensively destructive of human politics and just as intractable. Technics, as Winner wrote, was conceived of by the thinkers in this book as 'out of control', but this does not necessarily refer to a chaotic process: technics may be understood as out of (human or political) control but it is increasingly 'in control' as an independent process. That is to say, technology follows a pattern or mode of action that prioritises efficiency, effectiveness and the advancement of technology itself, which is dehumanising in many respects and which inevitably concludes with the domination of technology over human freedom and politics. These structural – political, social, psychological and material – transformations may not look 'catastrophic' in the moment, but it is with the unperceived catastrophe that is unfolding through these technological influences that these thinkers are just as concerned, as well as with the technological products that may emerge from these processes of technological evolution and acceleration. Equally, while technological 'catastrophe' need not be chaotic, it also does not necessarily model the master–slave dialectic; its 'domination' is more insidious – technology is clearly not of the same order of life as human beings, nor does it possess agency as humans do. It is closer, rather, to Arendt's model of bureaucratic rule (and shares much with ideas of bureaucratic rationality), whereby technology 'governs' through the types of organisation it effects on politics and society. 'Modern megatechnology', writes Jonas, contains the threat of:

> physical annihilation and that of existential impoverishment: the former by means of its unquestionably negative potential for catastrophe (such as atomic war), the latter by means of its positive potential for manipulation. Examples of this manipulation, which can lead to our ethical powerlessness, are the automation of all work, psychological and biological behavior control, various forms of totalitarianism, and – probably most dangerous of all – the genetic reshaping of our nature.[41]

What makes the problem of technology even greater in the writings of these post–Second World War/Cold War theorists is, in contrast with the 'apocalyptic' interwar texts by writers such as Walter Benjamin or Ernst Bloch, as Anson Rabinbach writes, 'the absence of any figure of redemption'.[42] Neither individual action nor existing political systems offer a solution for the transformation of this modern predicament. This is why, among the theorists who form the focus of this work, it is Marcuse and Heidegger, hailing from opposite sides of the ideological spectrum, who were most optimistic about technology's future, because they believed in the possibility of total revolution.

Outline of Chapters

Chapters 1 and 2 of this book focus on situating the critique of technology within the relevant historical contexts. Chapter 1 sets out some of the key personal, political and intellectual contexts in which the critique of technology developed, and the relationships and intellectual influences at play, looking at how we can understand this disparate collection of thinkers as a group, in relation to their ideas of technology specifically. Chapter 2 outlines the overlapping historical narratives of the development of technology that can be found in the work of the technological sceptics, and thus the idea of 'technology' as an overarching concept in itself. It shows how the theorists share a narrative about the distinctive position of technology in modernity: the claim that modernity is inexorably technological.

The following chapters focus, in turn, on specific forms of technology and technological practice: technologies of war, of production and work, media and communications, and biological and medical technologies. These chapters aim to show how the broader, more abstract concept of technology shaped, and reflected, understandings of the workings of individual technologies, and their influence on society and politics. 'Technology' was not only understood as a totality by its critics; particular technologies were also analysed in terms of their specific effects and the threats they posed to politics. Chapter 3 explores technologies of war, particularly nuclear weapons, outlining the relationship that the critics of technology saw between totalitarianism and modern technology. Chapter 4 looks at technologies of production, particularly the rise of autonomous technologies and machines, and how the critics of technology saw this transforming the structure of our lives and our relationships with almost every aspect of the material world, causing us to become increasingly alienated from worldly things, from human communities and from ourselves. Chapter 5 concentrates on technologies of media and communication, outlining the way in which the critics believed such technologies were increasingly closing down spaces of freedom in political modernity, whether through direct means such as propaganda or the more insidious influence of mass standardisation or even the restructuring of 'private' and 'public' spaces. Chapter 6 explores various

19

technologies of the body, and the fears of the critics of technology that contemporary technology was increasingly turning humans into mere 'raw material'.

Chapter 7 assesses the ways that technologies are seen to be transforming the structure of the contemporary world: our concepts of worldliness and our relationship with 'world'. This chapter argues that technology is seen to be catastrophically unsettling humans' relationship with their world: technology is considered to have radically altered our relationship with the earth and with nature, and not for the better. This brings the critics of technology into the realm of early environmentalism. The theorists of technology found in the environmental threat a degree of proof of their catastrophising, and in radical environmental politics also a degree of hope for the future. For its turn, environmentalism found in the technology critique much that aligned with its own fears and concerns. The book concludes with an examination of this interaction and some suggestions on the broader influence that might be traced from the technology critique, including the idea that the technology critique might be considered as an early 'Anthropocenic' form of understanding, highlighting some ways in which we might consider the claims of the critics of technology to be useful in our reflections on technology in the twenty-first century.

NOTES

1. Martin Heidegger, 'The Question Concerning Technology', in *The Question Concerning Technology and Other Essays*, trans. William Lovitt (New York: Harper & Row, 1977), 4.
2. Hans Jonas, 'Seventeenth Century and After: The Meaning of Scientific and Technological Revolution', in *Philosophical Essays* (New York and Dresden: Atropos Press, 2010), 80.
3. Arthur M. Melzer, Jerry Weinberger and M. Richard Zinman (eds), 'Preface', in *Technology in the Western Political Tradition* (Ithaca and London: Cornell University Press, 1993), vii.
4. Don Ihde, *Heidegger's Technology: Postphenomenological Perspectives* (New York: Fordham University Press, 2010), 19–20.
5. George Grant, 'Thinking about Technology' [1987], *Communio* 28 (2001), 614.
6. Lewis Mumford, 'The Corruption of Liberalism', *The New Republic*, 29 April 1940.
7. Jacques Ellul, *The Technological Society*, trans. John Wilkinson (New York: Random House, 1964), 203–4.
8. Theodor W. Adorno and Max Horkheimer, *Dialectic of Enlightenment* [1944/47] (London: Verso, 1997), 134.
9. Duncan Bell, 'What is Liberalism?' *Political Theory* 42:6 (2014); Samuel Moyn, *Liberalism Against Itself: Cold War Intellectuals and the Making of Our Times* (New Haven and London: Yale University Press, 2023).
10. Jerry Weinberger, 'Liberal Democracy and the Problem of Technology', in *Democratic Theory and Technological Society*, ed. Richard B. Day, Ronald Beiner and Joseph Masciulli (Abingdon: Routledge, 1988), 126.

11. Langdon Winner, 'Technologies as Forms of Life', in *The Whale and the Reactor* (Chicago: University of Chicago Press, 1986), 6.
12. Langdon Winner, 'Do Artifacts Have Politics?', in *The Whale and the Reactor* (Chicago: University of Chicago Press, 1986), 22.
13. Grant, 'Thinking about Technology', 622–3.
14. Leo Marx, '"Technology": The Emergence of a Hazardous Concept', *Social Research* 64:3 (1997), 980.
15. Ibid., 982.
16. Ibid., 984.
17. Langdon Winner, *Autonomous Technology: Technics-out-of-Control as a Theme in Political Thought* (Cambridge, MA: The MIT Press, 1977); Andrew Feenberg, *Transforming Technology: A Critical Theory Revisited* (Oxford: Oxford University Press, 2002); Andrew Feenberg, *Technology, Modernity, and Democracy* (London, New York: Rowman & Littlefield, 2018).
18. Andrew Feenberg writes: 'The critical theorist Marcuse sketched an answer I have tried to develop in what I call a critical theory of technology. According to critical theory, the values embodied in technology are socially specific and are not adequately represented by such abstractions as efficiency or control. Technology can frame not just one way of life but many different possible ways of life, each of which determines a different choice of designs and a different range of technological mediation. Does this mean that technology is neutral, as instrumentalism believes? Not quite: modern societies must all aim at efficiency in those domains where they apply technology, but to claim that they can realize no other significant values besides efficiency is to overlook the tremendous social impact of differing design choices.' In Feenberg, *Technology, Modernity and Democracy*, 62–3.
19. Andrew Feenberg, 'Critical Evaluation of Heidegger and Borgmann', in *Philosophy of Technology: The Technological Condition – An Anthology*, ed. Robert C. Scharff and Val Dusek (Oxford: Blackwell, 2003), 327.
20. Feenberg, *Technology, Modernity and Democracy*, 127.
21. Winner, *Autonomous Technology*, 174.
22. Hans Jonas, 'Technology as a Subject for Ethics', *Social Research* 49:4 (1982), 985.
23. Weinberger, 'Liberal Democracy and the Problem of Technology', 125–6.
24. Jacques Ellul, *The Technological Society*, trans. John Wilkinson (New York: Random House, 1964), 6.
25. Benjamin Lazier, 'Earthrise; or, The Globalization of the World Picture', *The American Historical Review* 116:3 (2011), 606.
26. Duncan Kelly, *Politics and the Anthropocene* (Cambridge: Polity, 2019).
27. For example: Barry Commoner, *The Closing Circle* (New York: Alfred A. Knopf, 1971); Rachel Carson, *Silent Spring* (London: Hamish Hamilton, 1962).
28. Delanty and Harris, 'Critical Theory and the Question of Technology: The Frankfurt School Revisited', *Thesis Eleven* 166:1 (2021); Michael C. Behrent, 'Foucault and Technology', *History and Technology* 29:1 (2013), 58–9; Eric Schatzberg, *Technology: Critical History of a Concept* (Chicago: University of Chicago Press, 2018).
29. Stanley Aronowitz, *Science as Power: Discourse and Ideology in Modern Society* (Minneapolis: University of Minnesota Press, 1998), 6–7.

30. 'German' here refers primarily to biographical background rather than to a linguistically German corpus of work: many important texts were written and/or published initially in English, a point which emphasises the complex international context of the technology concept.
31. John Burrow, 'Intellectual History in English Academic Life', in *Palgrave Advances in Intellectual History*, ed. Richard Whatmore and Brian Young (Palgrave: Basingstoke, 2006), 22–3.
32. Thomas P. Hughes, *Human-Built World: How to Think about Technology and Culture* (Chicago: University of Chicago Press, 2004), 2.
33. Hannah Arendt, 'Introduction *into* Politics', *The Promise of Politics* (New York: Schocken Books, 2005), 154.
34. Günther Anders, *Die Antiquiertheit des Menschen*, vol. 2: *Über die Zerstörung des Lebens im Zeitalter der dritten industriellen Revolution* [1980] (Munich: Verlag C. H. Beck, 2018), 9.
35. Herbert Marcuse, *One-Dimensional Man* (Boston, MA: Beacon, 1964), xvi.
36. Benjamin Lazier, 'Earthrise; or, The Globalization of the World Picture', *The American Historical Review* 116:3 (2011), 604n.
37. Ernst Jünger, *The Worker* [1932] (Evanston: Northwestern University Press, 2017); Oswald Spengler, *Man and Technics: A Contribution to a Philosophy of Life* [1931] (Budapest: Arktos, 2015); John Dewey, *The Public and Its Problems* (London: G. Allen & Unwin, 1927); Stewart Brand, *Whole Earth Catalog* (1968–1972); Buckminster Fuller, *Operating Manual for Spaceship Earth* (Carbondale: Southern Illinois University Press, 1969); Harold Wilson, 'Labour and the Scientific Revolution', policy statement made to the Annual Conference of the Labour Party, Scarborough, 1963; Shulamith Firestone, *The Dialectic of Sex: The Case for Feminist Revolution* (New York: Bantam Books, 1971).
38. John McDermott, 'Technology: The Opiate of the Intellectuals', in *Technology and Man's Future*, ed. Albert H. Teich (New York: St. Martin's Press, 1981), 135.
39. Herbert Marcuse, *Soviet Marxism: A Critical Analysis* (London: Routledge and Kegan Paul, 1958), 6.
40. Antonio Y. Vázquez-Arroyo, 'How Not to Learn from Catastrophe: Habermas, Critical Theory, and the "Catastrophization" of Political Life', *Political Theory* 41:5 (2013), 747.
41. Hans Jonas, 'Towards an Ontological Grounding', in *Mortality and Morality: A Search for Good after Auschwitz*, ed. Lawrence Vogel (Evanston: Northwestern University Press, 1996), 108.
42. Anson Rabinbach, *In the Shadow of Catastrophe: German Intellectuals between Apocalypse and Enlightenment* (Berkeley: University of California Press, 1997), 10.

I

COLD WAR CRITICS OF TECHNOLOGY

INTRODUCTION

Of the many critics of technology writing in the Cold War period, this book foregrounds the thought of nine thinkers who identify modern technology as essentially catastrophic for society, politics and the human condition. Those nine – Theodor Adorno, Günther Anders, Hannah Arendt, Jacques Ellul, Martin Heidegger, Max Horkheimer, Hans Jonas, Herbert Marcuse and Lewis Mumford – are central to this critical discourse in different ways.

Heidegger and Ellul have long been considered foundational in the study of twentieth-century philosophy of technology, and Marcuse too, although to a lesser degree.[1] Horkheimer and Adorno's *Dialectic of Enlightenment*, the most famous Frankfurt School text (albeit arguably rather atypical of the Institut für Sozialforschung's output), positions technology at the centre of its analysis of modernity.[2] These are all theorists who are widely known for centring the critique of technology in their work, or for writing significant and influential works on technology. In contrast, Mumford is a little-known figure now, but was a well-known public intellectual and frequently cited thinker on technology for much of the twentieth century, including by other now much more famous political philosophers and theorists.[3] Günther Anders, Hannah Arendt's first husband and a peripheral member of the Frankfurt School, wrote widely and influentially on technology in the post-war period, particularly on the advent of the atomic age, and while he has received some attention from contemporary German academics, his work is little read in the anglophone world, in part because most of it has not been translated into English. Jonas's work gained attention after the publication of his 1979 *Imperative of Responsibility* – in Germany in particular, but also on a wider international scale – but that interest has since waned.[4] Arendt, on the other hand, remains extraordinarily popular and widely read, although, while technology features prominently in some of her major works, its significance has been recognised by relatively few commentators.[5]

Collectively, the writings of this group reveal how their largely cohesive concept of technology was understood as a catastrophic force for modernity. Many others in this period also express similar concerns about technology, or some of its features, even if they would not agree that technology is catastrophic in its entirety, for example: Daniel Bell, Hans Blumenberg, Albert Borgmann, Bernard Charbonneau, Michel Foucault, Pierre Francastel, Erich Fromm, Arnold Gehlen, Siegfried Giedion, George Grant, Jürgen Habermas, Ivan Illich, Karl Jaspers, Reinhart Koselleck, Marshall McLuhan or Leo Strauss to name some of the thinkers who worked on or around the technology critique for at least some parts of their careers. The nine thinkers who feature in this work, however, are centred because they offer extensive, unambiguous and considered reflections on 'catastrophic technology', and in many cases are considered to be among the most significant theorists in the history of the philosophy of technology.

In the fields of political philosophy and intellectual history, Arendt, Adorno, Heidegger, Horkheimer and Marcuse are all widely recognised as major thinkers of the twentieth century; within the history and philosophy of technology Ellul joins this group. An extensive literature exists on all of these theorists which, in many cases, also explores their individual works on technology. But the way in which an overlapping conception of technology joins their thought together in a distinct tradition in this era (and which also influences or is present in the thinking of numerous other social and political theorists, in addition to those just cited) has not for the most part been recognised, except in a partial way, by looking at the ideas of the Frankfurt School, or the ideas of Heidegger and his students, for example.[6] In what sense, then, might we suggest that these highly intellectually and politically disparate theorists form a group in relation to the technology critique? Leaving aside the very many differences between them, some of which will be outlined in this and following chapters, the primary connection *is* their engagement in a shared critique of technology, as this book will argue. But there are also other important points of overlap.

The first is the unique German–American intellectual exchange that takes place in the middle of the twentieth century, largely as a result of the exile of many Jewish intellectuals in the 1930s and 1940s. Clearly, theorists of German origin, who were educated in a German intellectual context, are central to this discourse: Arendt, Adorno, Horkheimer, Marcuse, Heidegger, Jonas, Anders. With the exception of Heidegger, all of these German theorists were also Jewish.

24

Despite having many Jewish students, Heidegger was, notoriously, a prominent Nazi sympathiser. However, this is just as much an American story. Lewis Mumford is no longer widely read, but in the early to mid-twentieth century was a major figure in American intellectual life and his work on technology was well known and extensively cited. Arendt, Anders, Jonas, Adorno, Horkheimer and Marcuse were all part of the Jewish intellectual migration to the United States that took place in the 1930s and 1940s (and, for Jonas, the 1950s, after a period of time spent in Palestine/Israel and then Canada). Arendt, Jonas and Marcuse made America their home, while Adorno and Horkheimer returned to Germany after the war, and Anders to Austria (the home of his second wife). Even the most vehement upholders of German intellectual traditions among this group found themselves increasingly integrated into American intellectual life, and their work became in part a response to the new – Americanised – world they found themselves in after the war. The work of these various thinkers became, in different ways, foundational or highly influential within the American intellectual arena, not least with respect to their ideas of technology, which has, however, not been generally understood to constitute a connecting theme *between* these diverse theorists. Amid these major figures in American academia, Jacques Ellul seems an outlier. In some ways this is true. Ellul lived and died in Bordeaux, the same city in which he was born; he wrote and published in French. However, Ellul's influence, particularly his critical work on technology, was greatest in America, and his writings on technology were both extensive and broadly read at the time – although, as for Mumford, rather less so now. He drew on the German intellectual tradition, as well as those of his French contemporaries, but was also heavily influenced by Mumford's writings, which he cites frequently in his work.

Another point of connection relates to the intellectual influences these thinkers drew upon, which constituted a vast array of philosophers, as one would expect, with some notable overlap between many of them – Kant, Kierkegaard, Jünger, Spengler, Benjamin and Veblen, to highlight a few key figures. But two particular methodological approaches dominated their work, those of phenomenology and Marxism. The Heideggerians – Heidegger himself, Arendt, Jonas, Anders and Marcuse – used phenomenological approaches in their work (in varying degrees), while the Marxist theorists of the Frankfurt School included here – Adorno, Horkheimer and Marcuse – along with Anders and Ellul, were all guided by a (critical) commitment to Marxist theory. Clearly, there is overlap between the two, particularly in the work of Marcuse and Anders, but even where a wholesale rejection of one or the other positions takes place, as it does in Arendt's vehement critique of Marxism, or Adorno's equally pugnacious rejection of Heidegger's phenomenology of being, its presence remains significant *through* these critiques.[7] Phenomenology and Marxism were both considered provocations to thought that, whatever one's opinion, could not be ignored. They are all central to the philosophical world these thinkers inhabit, whether critically or positively.

It should also be highlighted that, despite these significant methodological connections, the critics of technology fall into a range of disciplinary categories. Heidegger is a philosopher in the traditional mould, as was Hans Jonas, although he formulated his clearly phenomenological philosophy explicitly in opposition to certain facets of Heidegger's thought. The same point might be made about Günther Anders, although he also incorporated Marxist social theory into his work, and is sometimes counted among the critical theorists of the Frankfurt School. Arendt, eschewing philosophy in the 1930s largely *because* she perceived in Heidegger (and philosophy at large) a complete abdication of political responsibility, insisted she be referred to as a political theorist, in reflection of her rejection of philosophy in the abstract. Adorno, Horkheimer and Marcuse all combined the critical social theory of the Frankfurt School with elements of psychoanalysis, while Horkheimer also brought an interest in the history of political thought to the table, Adorno an engagement with aesthetics (particularly that of music), and Marcuse a deeper engagement in radical contemporary politics. Mumford hailed from a very different tradition: beginning with a lifelong interest in architecture and cities, his work spanned social theory, political theory, cultural history and the philosophy of technology, in a remarkable and long career. Ellul, trained in sociology, wrote as prolifically on theology as he did on social theory, and considered the two aspects of his work to be equally important in understanding the world.

Finally, all of these thinkers explain the character of the Cold War world in a fundamentally technological way, and in a manner consistently critical of the dominant Western liberal paradigm. This is despite the fact that the political and philosophical positions they hold are in conflict. Their political-ideological affiliations spanned the spectrum from the farthest right (Heidegger, especially in the 1930s and 1940s), to the farthest left (Marcuse, particularly in his period as a 'guru' of the New Left in the 1960s), the political opinions and positions of these theorists are radically opposed to one another, and in some cases hostile. Arendt, who refused to categorise herself as either a liberal or a conservative, is most commonly considered to have aligned herself with a civic republican ideology, while Adorno and Horkheimer, nominally Marxist theorists, remained clearly on the left, although they abandoned much of Marx's political ideology. Ellul described his political commitments as Christian anarchism, a pacifistic opposition to repressive political power, and in his view, 'the fullest and most serious form of socialism'.[8] Mumford expresses at least some broadly liberal ideals along with an adherence to moderate forms of socialism (although he claimed he had 'never been a Liberal') while being deeply critical of the failure of modernity to realise these ideals in political practice.[9] Jonas, although criticised by some for his seemingly quasi-authoritarian claims about modern politics, considered himself dedicated to authentically democratic freedoms, and was sharply critical of the utopian schemes of Marxist philosophers.[10]

The personal relationships and influences *between* the members of this group of technological sceptics are similarly complex. Marcuse, Adorno and Horkheimer worked together for many years, Arendt and Jonas were lifelong friends despite holding very different philosophical and political ideas. Anders and Arendt were married for a short time in the 1920s and 1930s, and both had an antagonistic relationship with the Frankfurt School. Arendt considered the school (especially Adorno and Horkheimer) partially responsible for her friend Walter Benjamin committing suicide on the Spanish border as he tried to escape Nazi Germany in 1940 – had he been invited to join the School, he could potentially have left Europe much earlier. Anders (who was also a cousin of Benjamin) had his habilitation rejected by Adorno, acting as examiner, in the 1920s, leading to the curtailment of his academic career within the German university system. Heidegger taught Marcuse, Anders, Arendt and Jonas, and Arendt and Heidegger, now famously, had also been intimately involved when Arendt was an eighteen-year-old student under his supervision. While his former students, in the main, reviled Heidegger after his turn to Nazism, compounded by his later refusal to apologise for his wartime actions, Arendt and Heidegger renewed their friendship after the war, with Arendt advising and assisting Heidegger in the American publication of his works. Mumford was a staff writer for *The New Yorker*, for which Arendt regularly wrote, and the two knew each other personally to at least some degree, as one warm exchange of correspondence from 1965 attests.[11] Ellul, on the other hand, remains outside this complex of relationships, remaining physically on the European side of the Atlantic, even as his work gained prominence in the United States, although he engages extensively with Arendt's work on revolution in his own *Autopsy of Revolution* and with Mumford's work on technology.

There is, therefore, no single line of connection between these thinkers, nor could they be considered a 'group' in a comprehensive sense, although there are key loci of intellectual influences, concerns, and a dense network of interactions between the theorists, including vehement, deep and well-documented antipathies between some of these individuals. However, it is the intention of this study to show that, specifically, their critique and idea of technology in modern politics *is* substantially shared, and that this is not an accidental convergence but rather the product of an interwoven set of theoretical concerns, historical contexts, intellectual influences and reactions to contemporary politics, emerging in and out of a primarily German–American background.

The careers of some of these thinkers preceded the Cold War. Heidegger, Mumford and Horkheimer produced substantial work from the 1920s, and Adorno and Marcuse began publishing in the 1930s; for all, significant formative periods of their life predate the Cold War. But, in this work, the period of the Cold War and the narrative of technology expressed by these thinkers during that era will mainly be foregrounded. This is for two reasons. First, this study will make the claim that the narrative of technology that emerges among these thinkers in this period is substantially influenced by the experience of the Second World War and the Holocaust. Second, the narrative of technology explored here with specific reference to the political effects of contemporary technology is developed in and against the particular geopolitical context of the Cold War and (for these Western authors) the political narratives of a dominant liberal-democratic-capitalist ideological framework.

All of these thinkers are profoundly sceptical about the historical influence of technology on modern society and politics, and can be considered to be taking part in a cross-cutting discourse on technology, through which appears a conceptualisation of technology as an existential threat. The critique of technology, or the concept of technology as catastrophic, is fundamentally tied to a political critique of the Cold War Western liberal world, and neither technology nor their critique of modern politics can be fully understood without reference to the other. They thereby reject both the dichotomy of East and West, or communist and democratic worlds while also questioning the claim to 'progress' that Western liberal politics makes.

BIOGRAPHICAL INTRODUCTIONS

Heidegger, Mumford and Ellul: three key figures in the philosophy of technology

In the field of the philosophy of technology, Heidegger's influence is greater than any other single thinker. Yet, although Heidegger was a giant of European philosophy from the 1920s onwards, Mumford's published work on technology preceded that of Heidegger, and it was Mumford who was a greater influence on the critique of technology that would be developed in the Cold War period. Mumford's 1934 *Technics and Civilization* was widely read and cited, while Heidegger's best-known work on the subject would not be written until the late 1940s and 1950s, most significantly his 1954 essay 'The Question Concerning Technology'. For both Mumford and Heidegger, their thinking on technology and its relationship with politics or society can be traced back to the 1930s, although the events of the Second World War and the advent of the Cold War caused a marked shift in the work of both thinkers, from their different vantage points across the Atlantic: Mumford lost the earlier optimism he had about technology's likely transformation in modernity, while Heidegger's thinking on technology was catalysed by these events.

Ellul, although younger than both Mumford and Heidegger, also became interested in questions of technology in the 1930s, and this would remain the primary focus of his work throughout his career. Many have accused Ellul, accordingly, of trying to explain *everything* through technology.[12] One 1966 critic wrote that Ellul displayed a 'paranoid style', akin to works 'such as Rachel Carson's *Silent Spring*'.[13] Ellul spent his life and his career in France, yet many of his works went unread in his native country, while he was and is most widely read in the United States.[14] A prolific author, his most popular work was *The Technological Society*, published in French in 1954 and translated into English ten years later. Its publication won him his American success.[15] It 'earned him notoriety on both sides of the Atlantic and placed him among the ranks of important European and American voices – including Martin Heidegger and Lewis Mumford – interrogating the role of technology in twentieth-century life'.[16] The 'cultural upheaval' of the 1960s caused 'Ellul's voice as prophetic critic of bureaucracy' to be heard, writes David Lovekin.[17] In the history of the philosophy of technology, the work of Heidegger, Mumford and Ellul has been among the most influential of the twentieth century.

Elected to a professorship at Marburg in 1923, Heidegger quickly began to draw a following, which would include Arendt, Anders, Jonas and Marcuse. His most important work, *Being and Time*, was published in 1927. As the National Socialists gained ground, he became an apparently enthusiastic supporter, to the dismay of many of his former students, and in 1933 was appointed rector of the University of Freiberg by Hitler. Heidegger resigned from this position the following year yet remained a party member until the end of the war. He later sought to downplay his involvement in the regime, although he refused to apologise for the part he did play. In the 1930s, Richard Polt writes, Heidegger considered the Germans the 'metaphysical people', possessing 'great potential for a new confrontation with the question of being . . . [but] caught between Russia and America, which are slaves of the technological worldview'.[18] Yet in his notorious wartime notebooks, Polt writes, in 'passage after passage, Heidegger portrays machination as a global counter-essence, an understanding of being that dominates all contemporary systems and events, leaving us little of no hope or extrication'.[19] The war was seen as an inevitable expression of this machination.

Lewis Mumford, who was born and lived in New York, was primarily known for his writings on architecture and cities as *The New Yorker*'s architectural critic for over thirty years. He was also widely known for his work on technology. Although he held visiting professorships and fellowships at a number of institutions (Dartmouth, Berkeley, MIT, Harvard, Stanford, among others), he was not a career academic, but rather a writer.[20] Technology featured as a theme in his work from his earliest writings in the 1920s, although it was his 1934 *Technics and Civilization* that first focused entirely on the subject. This was the first work, its 1963 introduction claims, that revealed the interplay between social milieu and the progression of technology itself.[21]

This 'seminal' work, argues Stephen Sheehan, 'furnished one of the clearest articulations of this time on the cultural role and history of technology, and of technology as a totalizing concept'.[22] But while he described the effects of technology in the darkest possible terms, Mumford believed that humanity was on the brink of a more humanistic technology. His optimism rapidly faded. Mumford had close links with Germany in the 1920s. He associated with a number of German architects and city planners, and was 'particularly attracted to the social reform movements within the Weimar Republic', also spending time in Munich in the 1920s and early 1930s when undertaking research for *Technics and Civilization* at the Deutsches Museum. In turn, translations of Mumford's work received a warm reception in Germany.[23] The collapse of German civilisation into Nazi fascism was thus a shock that Mumford felt deeply.

In his wartime and post-war writings on politics and technology, however, it was not just German civilisation at stake, but civilisation itself. 'In a 1940 draft proposal for a new journal,' Heinz Tschachler tells us, 'Mumford declared that "the barbarian lies within us, no less than in the breast of the barbarian leader."'[24] In assuming the principles of total war, America had adopted the approach of its enemies. Later, in 1958, Mumford wrote of the essentially 'totalitarian notion of total war' by which America had '*lost the moral distinction that had hitherto separated ourselves from our totalitarian enemies*'.[25] Mumford's opposition to Nazi Germany was matched, Tschachler explains, by his objection to liberals such as John Dewey or Charles Beard who he accused of passivism and isolationism.[26] These later writings also reveal, Robert Casillo writes, a '"decisive devaluation" of technology's part in shaping society'.[27]

Born in Bordeaux in 1912, Ellul studied at Bordeaux and Paris. His key influences as a young man were Marx and Kierkegaard, and from 1932, when Ellul 'underwent a radical conversion to Christianity', the work of Karl Barth.[28] Initially, he explored Catholicism, but 'he found Protestantism closer to his understanding of the Bible, concerned more with a reading and interpreting of the text'.[29] His conversion to Christianity as a teenager continued to play a core part in his philosophical values and interests through his life: Ellul was a theological thinker as much as a sociologist or political theorist. During the 1930s, Ellul became interested in technology, working alongside Bernard Charbonneau to develop his idea of 'technique' as the ideology of technology.[30] 'It was Ellul's experiences leading up to and during the war', writes Michael Morelli, 'which gave ground to his worries about *technology's* idolization of *technique*.'[31] During the war, Ellul remained in France, although he was removed from his professorship at Strasbourg for his opposition to the Vichy Regime, while his father was imprisoned (ultimately dying in prison) and his wife forced to flee to rural France, in both cases because they possessed British citizenship.[32] At this time, Ellul became a member of the Resistance, helping the persecuted to escape to unoccupied France.[33] He was later awarded the title of Righteous Among the Nations by Yad Vashem for these activities. After the war, Ellul returned to teaching, this time in Bordeaux, where he would remain for the rest of his life.[34]

Heidegger's most developed thinking on the topic of technology, and by far his most influential work on the subject, can be found in 'The Question Concerning Technology'. 'Despite the bad odor of his politics,' argues Jerry Weinberger:

> Heidegger's argument about technology has greatly determined the course of modern thought in general, and on our shores it has influenced recent interpretations of liberal politics. Thus we cannot ignore Heidegger's claim that as technology stamps our age both technology and triumphant liberalism fall into crisis.[35]

However, Heidegger's work on technology comes *after* his former acolytes had broken away from him, and thus, with the exception of Arendt, who rekindled a friendship with Heidegger in 1950, it seems unlikely that Heidegger's work on technology had much influence over the other critics of technology in this work. In the 'Question', Heidegger characterises technology as 'enframing': via technology we come to reposition nature and world as 'standing-reserve', which in turn blocks our understanding of authentic being. 'The essence of technology is by no means anything technological,' he writes.[36] Technology cannot be reduced to its objects, nor even its processes. The nature of technology, he concludes, 'is a way of revealing'.[37] Despite the highly abstracted and conceptual notion of technology offered here, Andrew Mitchell has pointed out that Heidegger's work on technology is grounded in an understanding of, and concern with, machines themselves, pointing to his notebooks of the late 1940s and early 1950s in which he displays interest in specific industrial machines.[38] Against this, a German pastoral idyll is contrasted: the free-flowing Rhine as it was prior to the hydroelectric plant, the snow-covered mountains of the Black Forest that Heidegger would retreat to, the pre-modern peasant who still 'take[s] care of' the earth he cultivates.[39]

Mumford's post-war works on technology, of which the most substantial was the two-volume *The Myth of the Machine*, published in 1967 and 1970, reflects a deep cynicism about modern society and the technology that has shaped it so extensively. Modern technology, fundamentally dehumanising and repressive, has its archetype in an ancient form of social organisation that Mumford calls the 'megamachine': a machine made of human, rather than mechanical, parts. He gives the example of ancient Egyptian pyramid building as an early megamachine. 'The invention of the archetypal machine . . .

proved in fact to be the earliest working model for all later complex machines, although the emphasis slowly shifted from the human operatives to the more reliable mechanical parts,' he claims.[40] Modern society has seen the renewal and advancement of the megamachine since the eighteenth century: the rise of modern technology goes hand in hand with the rise of a technologised, and thus dehumanised, civilisation. Against this trend, there is little hope but, Mumford suggests, 'if we are to prevent megatechnics from further controlling and deforming every aspect of human culture', we might – and must – look to organic models.[41] Against mechanism Mumford pits a romantically inspired organicism. With these later works, writes Paul Forman, Mumford acquired 'an enthusiastic popular, and especially youthful following. Mumford must be placed with Herbert Marcuse as a major influence on the insurgents of the late 1960s and early 1970s.'[42]

While Ellul's interest in technology dates from the 1930s, he begins to write and publish extensively on technology only after he returns to academia after the war, particularly in *The Technological Society*. The central influence of Mumford is evident in his writings; *The Technological Society* frequently references Mumford's *Technics and Civilization* and *The Culture of Cities* (1938), albeit often critically. This is perhaps unsurprising, given that these works reflected Mumford's earlier quasi-optimism over humanity's technological future, an optimism which Ellul had never shared, and which Mumford himself had long moved away from by the time of the publication of *The Technological Society*. Ellul's analysis of technology refers to *technique*: that is, he is explicit that not technology in its particularity but the whole body of technology and its logic is his concern. (One critique he makes of Mumford is that he thinks too much in terms of 'the machine' rather than the extensive set of social processes that technology describes.) While 'technique certainly began with the machine . . . [it] has now become almost completely independent of the machine, which has lagged far behind its offspring.'[43] Technique, he writes, 'does not mean machines, technology, or this or that procedure for attaining an end. In our technological society, *technique is the totality of methods rationally arrived at and having absolute efficiency . . . in every* field of human activity.'[44] In *The Presence of the Kingdom*, Ellul describes technique as the subsuming of ends by means. '[T]oday everything has become means,' he writes. 'The end no longer exists . . . We have forgotten our common purposes, we have enormous means at our disposal, and we put into operation prodigious machines in order to arrive nowhere.'[45] Technique is basically limiting because 'what constrains us is that we no longer conceive of action except in the rational form of mechanical means'.[46] As a theorist who believed in the necessity of dialectical movement, the elision of end and means brings about a condition of stasis in social progress.

Ellul insists he is not opposed to technical work per se, but that:

it has no meaning if it is not guided, accompanied, and sustained by another work, one that Christians alone can do and yet often do not. For the world must be preserved by the ways of God and not by the technique of human beings (although technique can enter into the ways of God if we take care to hold it under judgment and submission).[47]

Technique, in the form in which it appears in modernity, is the problem then, not technology/technique itself. And a key issue is – not precisely secularisation (but connected to it) – the *sacralisation* of society. 'Man's fundamental experience today is with the technical milieu (technology having ceased to be mediation and having become man's milieu) and with society.' Yet this milieu is no more fathomable than man's environment has ever been: 'He needs axes of comprehension, of interpretation, of the possibility for action – that is, the sacred. Thus the desacralisation of nature, of the cosmos, and of the traditional objects of religion is accompanied by a sacralisation of society as a result of technology.'[48] Alongside the fact that our whole society becomes increasingly identical to technology and technique, secularisation (and the influence of science on the desacralisation of 'nature' seems key here) prevents us from seeking recourse to traditional ways of making sense of the world. Technique, the principle of technical society, itself takes on the role of the 'sacred'. Ellul critically draws on Marx (and Marxism, citing Stalin) to make this point: 'Technology is the instrument of liberation for the proletariat. It need only progress for the proletariat to free itself a little more from its chains . . . This is indeed a belief in the sacred. Technology is the god who saves.'[49] The bourgeois class in general have less faith in this, but 'the technicians of the bourgeois class are without doubt more strongly infatuated'.[50]

Horkheimer, Adorno and Marcuse: technology and the Frankfurt School

Many of the social theorists associated with the Institut für Sozialforschung, or, more popularly, the Frankfurt School, had an interest in science and technology's role in modernity from the start, although this interest flourished and took a particular shape in the era of the Second World War and the period following the war. Three of the Institute's most significant figures, Horkheimer, Adorno and Marcuse, wrote extensively on the perils of contemporary technology for modernity. The Frankfurt School's social critique lay between its rejection of German fascism and its concern with American capitalism: the politics of the Institute's two 'homes'. The Institute was founded in Germany in 1923 for the study of Marxist social and political theory, but some of its most significant work was undertaken in its temporary home in the United States between 1935 and 1949. Its influence on American intellectual culture was also substantial. Horkheimer and Adorno were 'prominent figures in scholarly and broader public discussions in the 1950s and early 1960s', although critical theory itself 'remained relatively unknown in the United States prior to Herbert Marcuse's appearances in public debates in the late 1960s'.[51] Yet it did have a substantial influence on many areas of intellectual life, among which, Martin Jay argues, its critique of mass culture and work on the potential for authoritarianism in America 'had the greatest impact on American intellectual life . . . For the first time, popular culture was attacked from a radical rather than a conservative direction.'[52] It spoke to, Jay argues, growing fears in America about the effects of mass democracy that could be dated back to Tocqueville, a century earlier.[53]

It was Horkheimer, first and foremost, who set the direction of the Institute, and who ensured its survival. Achieving his dissertation and habilitation at Frankfurt University in the 1920s, by 1930 Max Horkheimer was promoted to professor of philosophy at Frankfurt and, in the same year, elected director of the Institut für Sozialforschung. As Horkheimer took over it was becoming clear that the future of the Institute could not be guaranteed in an increasingly Nazi-dominated Germany. He began to make plans to move it out of Germany, plans which came to fruition in 1933. Moving first to Geneva, and in 1935 to New York, the Institute found a temporary home at Columbia University, where it would be based until it finally moved back to Germany in 1949. Horkheimer's first period at the helm, from 1930 to 1941 (he held the directorship again from 1950 to 1958), oversaw significant intellectual production, followed by his and Adorno's own masterpiece, the *Dialectic of Enlightenment*, written in the early 1940s.

Adorno's interest in politics and philosophy began at a young age, although it was always mixed with an interest in music, at which he had been precociously talented. Born in Frankfurt to an assimilated German-Jewish father, a successful businessman, and a Catholic Italian mother, a professional singer before her marriage, Adorno was surrounded with music from early on. It was music, then aesthetic theory, that constituted his first interests, and he came to social theory somewhat later, retaining his interests in music and aesthetics throughout his career. In his analysis of modernity – including the role of technology – Adorno is primarily concerned with the status and character of culture in modernity, which has been changed and denuded by modern technologies, particularly of media and production. Shortly after Horkheimer took up his directorship in 1930, Adorno began to write for the *Zeitschrift für Sozialforschung*, the journal of the Institute. In 1933, Adorno left Germany for Oxford, where he stayed for four years, travelling to New York in 1937 after Horkheimer extended him an offer to join the Institute, and later moving with Horkheimer to California to work with him on the *Dialectic*.

In contrast to Horkheimer and Adorno, Marcuse's engagement with politics began in practice. He was drafted into the German Army as a young man during the First World War, and took part in the German Revolution of 1918–19 as a member of the soldier councils. The experience of war and revolution was foundational for Marcuse; its impact would stay with him throughout his life, and the sense of possibility afforded by the revolution.[54] However, while he identified as a Marxist, he became disillusioned with party politics in Germany, and profoundly concerned with the development of communism in Soviet Russia, which led him early on to keep politics at arm's length for much of his career. Following the war, Marcuse pursued a career in academia. He was deeply influenced by Heidegger, who supervised his habilitation between 1928 and 1932, and he sought to combine Heidegger's work with Hegelian and Marxist philosophy. In 1932, Marcuse was hired by the Frankfurt School. Moving with the Institute, Marcuse worked briefly in Geneva and Paris before emigrating to America, where he would remain until his death in 1979. In 1940, Marcuse became a US citizen and spent time in the 1940s working for the Office of War Information and the Office of Strategic Services in Washington, DC, providing intelligence reports on Nazi Germany and Soviet Russia.

Horkheimer's early work in the 1920s shows an optimism about the role of science and technology in society that echoes Marx. He did not criticise the attempted mastery of nature, as he would famously do later, but instead held a more orthodox Marxist position, Wolf Schäfer explains, that 'classless society will unleash the powers of science and technology and achieve complete human mastery of history'.[55] But a 'reversal of premises' and 'turn to a pessimistic disposition' is clear by the time he writes his 1942 article 'The End of Reason'.[56] Horkheimer's engagement with technology is largely through his questioning of the idea of 'reason', a theme that, in different ways, is central to his intellectual project. This takes the form of a historical analysis and critique of science and the 'scientific attitude', science's relationship with technology, and the influence of both on autonomy, freedom and the individual in society.[57] The Enlightenment sought to make rationality the principle that underpins society, for the betterment of man, but Horkheimer suggests the concept contains 'defects that vitiate it essentially'. The Enlightenment sought the progressive expansion of reason, and the result of this was the growth of the economy (the Industrial Revolution), and the development of technical means, including techniques and bureaucracies as well as machines. Man's horizons were expanded, and his breadth of activity grew, but:

> his autonomy as an individual, his ability to resist the growing apparatus of mass manipulation, his power of imagination, his independent judgment, appear to be reduced. Advance in technical facilities for enlightenment is accompanied by a process of dehumanization. Thus progress threatens to nullify the very goal it is supposed to realize – the idea of man.[58]

Rationality, while initially taking the form of an attempted mastery of 'nature', eventually comes to mean (or, rather, is unmasked as) the domination of all men. Reason becomes 'the mere instrument of the all-inclusive economic apparatus'.[59]

In the *Dialectic of Enlightenment*, Horkheimer's work on reason and his historical reading of the Enlightenment came together with Adorno's critique of the culture industry. For both, technology plays a key role in modernity, comprehended *as* domination. Adorno believed that the culture industry, and the monopoly and totalisation that essentially characterise that industry today, is a precursor to how other sectors of the economy and society will function in the future, as well as playing an active role in bringing about a 'total' society. These changes have brought about the increasing dehumanisation of modern humans, a dehumanisation they are not even aware of due to the promotion of 'needs' to feed economic expansion, and to the 'propaganda' of modern culture. Adorno explicitly makes a case for the equivalence of technologies: that all technology is a unitary or unifying mechanism, and this is clearest in the way he argues that the culture industry and its technologies are merely an outgrowth of the technologies of the production sphere. 'Automobiles, bombs, and movies' keep society together, he writes.[60] The technologisation that has already occurred in the culture industry will spread to the rest of society: 'The ruthless unity in the culture industry is evidence of what will happen in politics.'[61] Furthermore, it will spread to the rest of society *via* the culture industry, as the 'whole world is made to pass through the filter of the culture industry'.[62]

This equivalence of totality brings forth an apocalypse for society. 'The more total society becomes, the greater the reification of the mind and the more paradoxical its effort to escape reification on its own. Even the most extreme consciousness of doom threatens to degenerate into idle chatter.'[63] This, as for other thinkers, is clearly an outgrowth of Adorno's reflections on the Nazis and the Holocaust, which is read as an elimination of negation. 'Genocide is the absolute integration. It is on its way wherever men are leveled off – "polished off," as the German military called it – until one exterminates them literally, as deviations from the concept of their total nullity. Auschwitz confirmed the philosopher of pure identity as death.'[64] For Adorno, technological power is very clearly derived from economic power, but takes the economic logic of domination further than ever before. The 'basis on which technology acquires power over society is the power of those whose economic hold over society is greatest. A technological rationale is the rationale of domination itself. It is the coercive nature of society alienated from itself.'[65] That is, technology *itself* takes the place of the elite class – a wholly dehumanised domination.

After the war, Marcuse returned to teaching, at Columbia, Brandeis, Harvard and San Diego. He published major works including *Eros and Civilization, Soviet Marxism* and, most influentially, *One-Dimensional Man*. He also published numerous articles and essays. In the late 1960s, Marcuse came to be considered something of a guru or figurehead for the New Left. This was not without its strains. Nonetheless, as Richard Wolin explains, 'in 1968 a contemporary observer could describe Marcuse as "the most widely discussed thinker within the American Left today."'[66] Marcuse, for his part, although he was by no means blind to the limitations of the New Left, saw in the movement a possibility for revolutionary political change – or, at least, a move towards authentic change – that was largely absent in modernity. His relative optimism, in fact, places him at odds with many of the theorists in this work, including his Frankfurt School colleagues.

This optimism extended to the topic of technology, where he expressed belief that the oppressiveness of technological modernity just might be overcome. Yet he was no less convinced of the oppressive nature of modern technology than any of the other theorists mentioned here. In an essay from the early 1940s, Marcuse writes of how technological rationality has become all consuming, serving only itself. 'The facts directing man's thought and action are not those of nature . . . or those of society . . . Rather are they those of the machine process, which itself appears as the embodiment of rationality and expediency.'[67] This rationality creates its own values, 'which hold good for the functioning of the apparatus – and for that alone'.[68] Later, in *One-Dimensional Man*, Marcuse develops these ideas, arguing that technology was, as such, fundamentally totalitarian in nature.[69]

Jonas, Arendt, Anders and the nihilism of technology

Jonas, Arendt and Anders first became friends as students at Marburg in the mid-1920s, where they all studied, for a time, under Heidegger's supervision. Arendt and Anders were married in 1929; they would divorce amicably in 1936. While Anders and Arendt had little to do with one another after that time, Arendt and Jonas maintained a lifelong friendship. Both Jonas and Arendt, argues Richard Wolin – and the same can be said of Anders – 'perceived a Faustian-nihilistic strain in Western humanism – the loss of a sense of proportion and "limit" – that seemingly propelled the modern age into the abyss'. The nihilism exposed by the Nazis, for them, 'lived on in the manifestations of modern technology: the risks of nuclear annihilation, environmental catastrophe, and interplanetary disorientation'.[70] This sense of nihilism was also at the heart of their mutual critique of Heidegger. But although motivated by similar concerns, and sharing an intellectual background, the three took quite different disciplinary approaches in their work, and their arguments often diverged from or even opposed one another's. Arendt, the political theorist, argued that politics *must* not be understood in primarily ethical terms, but rather as a unique activity in its own right, whereas Jonas, primarily a philosopher of ethics, would later in his career come to make political arguments based on his ethical philosophy. Anders, on the other hand, described his work as 'philosophical anthropology', and unlike either Jonas or Arendt was significantly influenced by Marxism and critical theory. His work focused above all on the influence of nuclear power on modernity, although this incorporated a wide-ranging analysis of technology more broadly. 'He was one of the first thinkers to conceptualise the nuclear age as a distinct world-historical epoch where the future of humankind as a whole had become precarious,' writes Jason Dawsey.[71] The three also propose contrasting and distinctive notions of the relationship between nature and technology, which all nonetheless emphasise the corruption of nature through means of modern technology.

Hans Jonas is best known for *The Imperative of Responsibility*, an influential text for the green movement in the 1980s, particularly in Germany.[72] Here, he makes the case that technology has transformed human action such that a wholly new kind of ethics is required in order to resist the dangers emanating from technological change. Although he began his intellectual career with a study of Gnosticism, an early and heretical Christian creed, for most of his career Jonas was a philosopher of ethics, although one who increasingly found himself engaged in political questions. Born in Germany in 1903, Jonas studied at Freiberg, Berlin and Heidelberg, gaining his doctorate on Gnosticism at Marburg with Heidegger. After Hitler rose to power, Jonas, a Jew and active Zionist, moved briefly to England, and then, in 1934, to Palestine. He settled there until 1940, when he joined the British Army, fighting in Italy and Germany. He found, on returning to Germany after the war, that his mother had been among those killed in Auschwitz. After he discovered this, he returned to Palestine, vowing never to live in Germany again. He resumed his academic career: briefly teaching at the Hebrew University of Jerusalem, before emigrating to Canada in 1950, followed in 1955 by a move to New York, where he taught at the New School for over twenty years.

Of the theorists foregrounded in this study, Arendt is probably the most widely read today. Unlike for Heidegger and the Frankfurt School thinkers, the role of technology in her work has not generally been emphasised. Yet it is fundamental to her thinking on modernity. Her most systematic and arguably most important work, *The Human Condition*, begins with the observation that we live in a world transformed by technology, and it is in this context that the work is described as an attempt to 'think what we are doing'.[73] Raised in Königsberg, Arendt studied at Marburg with Heidegger and Heidelberg with Jaspers, leaving Germany in 1933 (following her arrest and brief imprisonment for research into antisemitism). She lived, until 1940, in Paris and in that period divorced Günther Anders, whom she had married in 1929, and married Heinrich Blücher, whom she met in Paris. In 1940, Arendt was again imprisoned, this time in the Gurs camp for enemy aliens. Amid the chaos of France's fall to Germany that took place soon after her arrival at the camp, she managed to escape, and she and her husband eventually secured visas to the United States, where they arrived in May 1941.

Both Anders and Blücher were, for periods of their life, affiliated with communism, Anders through his association with the Frankfurt School's Marxism, Blücher through his involvement in the Spartacist Rebellion in Germany in 1918–19. Arendt, though sympathetic to, for example, Rosa Luxemburg's writings, was critical of Marxist ideology as a whole. Her interest in politics began, rather, with her engagement with Zionism in the 1930s and 1940s, in which she was deeply invested although ultimately disappointed in what she considered to be its increasingly nationalistic character. The political ideology she came to associate with most was a kind of civic republicanism, drawing on the traditions of ancient Greece and Rome but also on the modern American republican constitution.

Günther Anders, while part of the intellectual migration to the United States in the 1930s, never attained the degree of success in the anglophone world that some of his fellow travellers did, and remains best known in his native Germany (only a limited part of his work has been translated into English to date). He was born Günther Stern, adopting his pseudonym in the 1930s when he worked as a journalist. He had studied under Husserl and also Heidegger, along with Arendt, Jonas and Marcuse. He was 'a participant in Weimar leftist circles that included Bertolt Brecht and George Grosz', writes Dawsey, as well as a cousin of Walter Benjamin. Although Anders was among the most devoted of Heidegger's followers, as the latter became closer to the Nazi Party, Anders distanced himself from his former teacher. At the same time, he began to study the early Marx, particularly his ideas of alienation, which became pivotal to his own thought.[74] He was on the periphery of the Frankfurt School circle, and while certainly influenced by their strain of Marxist social theory, as well as publishing in the *Zeitschrift für Sozialforschung*, he had also (as earlier noted) had his habilitation rejected by Adorno, and he was never formally in the employ of the Institute. As Christopher John Müller explains, '[H]is engagements with National Socialism shared a proximity with Marxism and the Frankfurt School of Critical Theory, but pursued their own course, and led to "a fragile existence" as a freelance author rather than an academic career.'[75] After the Reichstag fire in February 1933, he fled to Paris from Berlin, where he lived for three years, before moving to the United States where he lived between New York and California.[76]

When Jonas returned to academia after the war, he turned to the question of how one might rectify the (significant) mistakes he believed Heidegger had made in his work, primarily the ethical nihilism he believed was inherent in his philosophy. The experience of the Second World War forced Jonas to confront the realities of modernity, as he explains. While in the army:

> cut off from books and all the paraphernalia of research, I had to stop work on the Gnostic project perforce. But something more substantive and essential was involved. The apocalyptic state of things, the threatening

collapse of a world, the climactic crisis of civilization, the proximity of death, the stark nakedness to which all the issues of life were stripped, all these were ground enough to take a new look at the very foundations of our being.[77]

The war was a 'watershed for philosophy', he would later write.[78] This was true in a number of respects, not least that the war, and the emergence of nuclear power was a moment of revelation about humanity's relationship with technology. The shock of the nuclear explosions would wake Jonas to the systemic transformation that a wide range of technologies effected in society and politics. Over time, his exploration of this became a consciously political project, as well as a philosophical one.

Technology is transformational and revolutionary. This revolution has been going on for centuries, Jonas writes.[79] Today, modern technology is distinguished by its extraordinary scale, which qualitatively transforms politics and ethics. 'Modern technology has introduced actions of such novel scale, objects, and consequences that the framework of former ethics can no longer contain them,' he writes.[80] 'The containment of nearness and contemporaneity is gone, swept away by the spatial spread and time-span of the cause-effect trains which technological practice sets afoot, even when undertaken for proximate ends.'[81] The impact of technology is so great as to change the world itself; we now understand, he argues, 'the critical *vulnerability* of nature to man's technological intervention'.[82]

Technology thus causes a revolutionary transformation of society, politics and the physical reality of human beings. Yet, Jonas writes, this is allied with a disturbing belief that all this is for the best. 'Optimism, as confidence in man, in his powers and natural goodness, is the signature of modernity.' We utterly fail to comprehend 'the revenges which nature – human and environmental – may hold in store, in the immense complexity of things and the unfathomable abyss of the heart, for the planner of radical change'.[83] We have become utopian in our thoughts, he writes, and the scale of modern technology makes former utopias seem viable.[84]

These ideas are expressed and developed most comprehensively in *The Imperative of Responsibility*. A new ethical imperative is required in this technological age, he writes, because 'with certain developments of our powers the *nature of human action* has changed'.[85] The ethic Jonas is seeking is explicitly 'anti-utopian'.[86] The critique of utopia he offers is 'a critique of technology in the anticipation of its extreme possibilities'.[87] It is modern technology *combined with* its utopian, progressive ideals that Jonas truly fears. The atom bomb, appalling though it is, has a degree of predictability about it – that is, we know the devastation it will wreak if used. Jonas's greater concern was the total unpredictability of the consequences of the use of nuclear energy even for peaceful, and notionally benign, use.

In her work, Arendt highlights various technological events in her lifetime – such as the birth of the atomic bomb and the dawn of space travel – that she sees as the culmination of a transformational and alarming technological development that first emerged in the sixteenth century. The key terms through which we can understand Arendt's history and critique of technology in the modern age (and through history) are 'world alienation' and 'earth alienation'. Both of these terms are associated with a scientific and technological turning away from the human, intersubjectively constituted world of experience and action, into a world understood in an objective and instrumentalised manner. Although 'instrumentalised', however, this also marks a turning *towards* nature and away from the constructed artifice of human politics ('world'), basing politics upon the patterns of nature, an assumption that a 'natural' politics is the best form, and even an acting *into* nature, or creating a new kind of nature. All these things, however, replace the true form of human artifice with a chaotic and erroneous – and dehumanised – pattern for human living that results inevitably in forms of domination.

In 1950, Anders moved back to Europe, not to Germany but to Vienna, the native city of his second wife, Elisabeth Freundlich. Like Jonas, he too asked the question of where Heidegger had gone so badly wrong, identifying, like Jonas, a fundamental ethical lacuna in his work. In the 1950s, Anders began to publish work on technology, gaining 'an international reputation as a critic of modern technology, an advocate of Holocaust remembrance and a tireless voice against the Bomb'.[88] While his thinking on technology was inspired by opposition to the nuclear threat, his understanding of technology and its social and political character extended beyond this. Mechanisation or technologisation described most activities, particularly labouring activities, in the modern era. Anders observed, writes Dawsey, that, in the 'push-button' era, 'the diminution of human agency had reached such an extent that, in wartime, the mere pressing of a button could quickly unleash devastation on the other side of the globe'.[89] Anders also wrote extensively on how technology transformed leisure and cultural activities, as the technological logic defining the labouring sphere of activity transcended the economic world to dominate all spaces of action, economic, social and political. The ethos of the 'push-button' era resides also, therefore, in the technology of broadcast television, where we are presented with images which deprive us of the *'the ability to distinguish reality from appearance . . . We become "passive."'*[90] These 'phantoms' or modern myths presented to us thus conceal any kind of authentic reality.

INTELLECTUAL INFLUENCES

Among this group, German thinkers dominate. Heidegger and his students Anders, Arendt and Jonas; from another intellectual tradition, Horkheimer, Adorno and Marcuse. The reason for their dominance in this study on sceptics of twentieth-century technology and its influence on politics can be traced back, in part, to a flourishing tradition of thinking about technology and its social influence in Germany – both positive and negative. Notably, Jeffrey Herf's work *Reactionary Modernism* has highlighted the way in which technology was reconciled with a neo-Romanticist ideology by conservatives such as Spengler and Jünger in the Weimar Republic, incorporating technology (but not materialism) into German *Kultur*, against Western *Zivilization* and its Enlightenment roots.[91]

It was not just German thinkers who were shaped by Weimar's intellectual climate, however. Mumford, whose mother had been the daughter of German immigrants, spent time in Germany in the 1920s, and his thought was influenced by both German thinkers as well as by American philosophy.[92] In one 1951 work, Mumford described the prescience of Spengler's analysis of civilisational breakdown in the aftermath of the First World War. 'Politically, our moral breakdown has taken precisely the turn predicted by Henry Adams fifty years ago, and by Oswald Spengler, with even more brutal realism, after the First World War.'[93] In post-war France, writes Tony Judt, an 'enthusiasm for modern German thought ... was now thoroughly incorporated into the indigenous French variant; among its central props was the Heideggerian distaste for "technical civilization"'.[94] Ellul's intellectual influences were broad and varied, but aside from the influences already mentioned there was a distinctly German engagement in his work.[95] Friedrich Jünger (the brother of Ernst) is frequently referenced in his works, as was the Swiss-German writer Siegfried Giedion, and he draws on the concept of reification as developed by Lukács and used extensively by the theorists of the Frankfurt School.[96] Mumford's work also exercised a clear influence on Ellul from his earliest writings.

Herf's study of Weimar's 'reactionary modernists' explores a 'cultural paradox ... the embrace of modern technology by German thinkers who rejected enlightenment reason'. In the period leading to the Nazi's seizure of power, he writes, 'an important current within conservative and subsequently Nazi ideology was a reconciliation between the anti-modernist, romantic, and the irrationalist ideas present in German nationalism and the most obvious manifestation of means-ends rationality, that is, modern technology'.[97] Herf includes among this group Ernst Jünger, Oswald Spengler, Werner Sombart, Hans Freyer, Carl Schmitt and, at one step removed, Martin Heidegger. The last, Herf argues, undertook a later shift in his attitude to technology, but in this period made a 'tenuous peace' with technology.[98] Naturally, not all Weimar-era thinkers, writers or philosophers could be considered reactionary modernists, and there were many on the right who continued to be opposed to the expanding influence of technology in society.[99] However, among the reactionary modernists, and arguably its flag-bearer, Ernst Jünger was a major influence on some of the thinkers in this study; Heidegger was another. This is not to suggest that the technological critique outlined here derived from the technological optimism of the Herf's group, but that in the early period of their careers, many of the thinkers in this book were steeped in an atmosphere of these kind of ideas. Michael Zimmerman observes that, for instance:

> Heidegger's later notion that in the technological era humanity is 'challenged forth' by *Gestell*, i.e., compelled to treat itself and other entities as standing-reserve (*Bestand*) for total mobilization, was analogous to Jünger's notion that the *Gestalt* of the worker compelled humanity to mobilize itself and the earth as standing-reserve for enhancing the technological project of total control.[100]

For Marcuse, Anders, Arendt and Jonas, Heidegger (in particular his work in the 1920s that culminated in *Being and Time*) was a major influence, and parallels and engagements between aspects of Heidegger's thought and that of various Frankfurt School theorists have been observed.[101] Mikko Immanen argues that a 'contestation with Heidegger's competing philosophical revolution played a considerable, unacknowledged role in the formation of the Frankfurt School . . . [Heidegger was] the most provocative challenge and competitor to their own analyses of the discontents of European modernity at the time.'[102] Wolin argues that Adorno borrowed 'too readily from the lexicon and habitudes of 1920s *Kulturkritik*', and Adorno's later (1964) work, *The Jargon of Authenticity*, is critically aimed at German existentialism as exemplified by Heidegger, in whose work, Adorno writes, '"Authenticity" has an "aura" of meaningfulness, whilst actually being absent of all meaning and infinitely malleable (and banal).'[103] During the last years of the Weimar Republic, writes Richard Wolin, 'Jaspers and Heidegger stood out as the "titans of existentialism"'. Like the Frankfurt School, Wolin argues, Heidegger internalised the 'dominant motifs of Weimar-era cultural criticism: above all, a fear that "culture" and "civilization" were mutually exclusive concepts; that civilization's rise went hand-in-hand with cultural decline'.[104] The same fears could be said to be reflected in the work of many of his other students, certainly Arendt and Jonas. However, the way each used Heideggerian ideas in their work, and their response to Heidegger's political moves in the 1930s, differed quite significantly. Marcuse, in the early 1930s, developed a synthesis of Heideggerian and Marxist ideas, although abandoned publication of this work following Heidegger's turn to Nazism, and the discovery of Marx's *Economic and Philosophic Manuscripts*. Marcuse's worldview would remain, however, Wolin argues, strongly influenced by German thought of the 1920s.[105] Jonas's post-war *The Phenomenon of Life* rejected the abstracted concept of Being that Heidegger had offered in *Being and Time*, arguing that, absent a conception of man as a biologically living being, Heidegger's philosophy collapsed into nihilism. Arendt's concept of natality, the genuinely free action of individuals creating anew, which underpinned her notion of the political, is clearly a direct refutation of Heidegger's reliance on mortality in his framing of the meaning of Being. And in the immediate post-war period, Anders focused primarily on developing a critique of the nihilism he identified in Heideggerian philosophy.

While all four split with Heidegger after his turn to Nazism, Arendt chose to reconcile with him in the years after the war. From 1950, the two exchanged correspondence and met sporadically, and Arendt became pivotal to the publication of Heidegger's work in America, corresponding and negotiating with editors on his behalf. Yet she continually expressed reservations about Heidegger (in her private correspondence to others, for example), and her engagement with his ideas was generally critical. Heidegger's own solutions to the ills of modern society have been informed by the scientific ideals of modernity, she intimates: 'If it does not belong to the concept of man that he inhabits the earth together with others of his kind, then all that remains for him is a mechanical reconciliation by which the atomized Selves are provided with a common ground that is essentially alien to their nature.'[106] As Yaqoob writes, in Arendt's eyes, Heidegger 'could grasp world-historical processes, but not the political character of the "world" that was in the process of being lost'.[107]

The other unavoidable influence on the work of all these thinkers was Marx, in quite distinctive ways. The Institut für Sozialforschung was set up to develop Marxist social theory: Horkheimer, Adorno and Marcuse were part of that enterprise, with Anders also within the same sphere of influence. Marxist theory was a pivotal influence on Ellul, even while he rejected aspects of Marx's thinking. On the other hand, Heidegger considered Marxism to be 'the most extreme nihilism', owing to its absolute prioritisation of the material and productive aspect of humanity; Arendt decried Marxist theory for similar reasons – because labour was positioned as the fundamental human activity – and Jonas's key antagonist in *The Imperative of Responsibility* was the Marxist (and Frankfurt School) theorist Ernst Bloch, whom Jonas denounces for his Marxian faith in revolutionary, progressive hope for the future.

At the time that Horkheimer took over the directorship of the Institut für Sozialforschung in 1930, as Martin Jay explains, 'social philosophy, as Horkheimer saw it, would not be a single *Wissenschaft* (science) in search of immutable truths. Rather, it was to be understood as a materialist theory enriched and supplemented by empirical work.'[108] This was the mission of the Frankfurt School under his leadership, and the philosophy and method behind their studies. As Jay writes, early Horkheimer staunchly defended:

> a dialectical notion of reason against the irrationalist alternative he identified with the threat of fascism. While acknowledging that rationality in its liberal, bourgeois form was tied to egoistic self-preservation

rather than the general good, he had resisted the conclusion that reason *tout court* inevitably led in this direction.[109]

Yet, between the 1930s and 1940s, the optimistic strain of Horkheimer's philosophy, that which had earlier reflected the promise of Marxist revolutionary principles and which even defended some aspects of Enlightenment reason, was increasingly eroded, in favour of his claims of the 'eclipse of reason' in modernity and even his move away from a more materialist historical approach. *Eclipse of Reason* and *Dialectic of Enlightenment* (both published in their final forms in 1947) represent 'the culmination of Horkheimer's increasingly sweeping critique of Western rationality', Abromeit writes, and an increasingly dehistoricised narrative.[110]

Adorno's distinctive contribution to Frankfurt School Marxism was his analysis of aesthetic theory, particularly as that evolved into a critique of the culture industry. Adorno first came to America to work on the Radio Research Project, a Rockefeller-funded exploration of the effects of mass media on society. In this project, his complex intellectual relationship to Marxist theory, and other schools of Marxist thought, can be seen. Adorno critiqued the 'deleterious effects of radio by pointing to its stimulus to standardization', explains Martin Jay. 'Although relating this to the permeation of the exchange ethic of capitalism, he also saw a connection with technological rationality itself . . . similar to Horkheimer's analysis of trends in the authoritarian state . . . "technical standardization leads to centralized administration."'[111] His concern with technology distinctively separated him from Leninist forms of Marxist aesthetic criticism, and its 'general indifference' to technology, writes Jay.[112]

In some respects, Marcuse remained the most 'orthodox' Marxist of the three Frankfurt School theorists in this study; in other ways, he was distinctly idiosyncratic. Unlike Adorno and Horkheimer, Marcuse retained a faith in the possibility of revolutionary change and did not fall prey to absolute disillusionment. He became sceptical about the agency of the proletariat, but maintained that, although Marx had been mistaken about the revolutionary process – and that modern society *did* tend to repress the dialectic – revolution was possible through the action of other groups who were genuinely 'outside' society. He maintained, to the end of his career, that he was an orthodox Marxist thinker.[113] Marx had simply been 'too optimistic and idealistic', he writes.[114] Importantly, what he had failed to understand was 'the great achievement of technological society: the assimilation of freedom and necessity, of satisfaction and repression, of the aspirations of politics, business, and the individual'.[115] As the producing classes have their desires (apparently) satisfied by the abundance of technological capitalism, what is needed is the emergence of 'social classes whose life is the very negation of humanity, and whose consciousness and practice are determined by the need to abrogate this condition'.[116] Thus, Marcuse concludes, 'the failure of humanism seems to be due to over-development rather than backwardness'.[117] That technological 'suppression is no longer terroristic but democratic, introjected, productive, and even satisfying' does not change the fact that it is in fact suppressive of political change and choice.[118]

As a young man of seventeen, in the wake of the 1929 crash, Jacques Ellul read, and was struck by the insights contained in, Marx's *Das Kapital*. Yet, 'Ellul disagreed with Marx's undialectical belief in the reality of progress and in the so-called liberating capacities of work.'[119] In 1932, Ellul underwent his radical conversion to Christianity. This conversion led to his rejection of Marxism in certain key respects. 'He found its political agenda, calls for violent revolution, rejection of religion, and lack of attention to human agency offensive,' writes Lovekin. 'But *he did not reject the critical agenda of Marx* – his attempt to offer a comprehensive critique of the predominant social order. Nor did he reject the dialectical approach that was characteristic of Marx.'[120]

Although Horkheimer, Adorno, Marcuse and Ellul all adopted, adapted and rejected aspects of Marx, Arendt, Jonas and Heidegger all emphatically rejected his ideas. Arendt, although it has been suggested that her husband, Blücher, the former Spartacist, had a more significant influence on her thought than is apparent in her texts, is dismissive of Marx's thought.[121] Marx committed the fatal theoretical error in his politics, she believed, the 'surrender of freedom to necessity', in emphasising material production as the foundation of society.[122] For his part, Jonas writes of how a 'fundamental error' in Marx 'is the separation of the realm of freedom from the realm of necessity, the belief that the one begins where the other ends, that freedom lies somewhere beyond necessity instead of in the meeting with it'.[123] All of these thinkers, however, rejected Marx's optimism about the potentiality of technology. This is the case even for Marcuse, who, while distinctly more optimistic about the chances for a humanistic, post-revolutionary technology, still associates contemporary society with its technological rationality to an extent that goes beyond Marx. Technology itself, as well as society as a whole, must be transformed in order to lose its repressive quality.

The influence of German philosophy on Mumford has been noted above. However, influence also travelled in the other direction. Like Heidegger, Mumford might be considered a key early writer of influence on many of the other critics of technology here. In almost every other way, the two could not be more different. Unlike Heidegger, a philosopher of the German academic elite and lauded as a 'genius' even as a young intellectual, Mumford did not even complete his university degree.[124] After a brief stint as a radio electrician in the US Navy in the final year of the First World War, he began his career in journalism, writing on literary criticism. Over his long career, his attention increasingly became focused on technological questions, and his outlook became increasingly bleak. Marcuse, in his 1941 essay 'Some Social

Implications of Modern Technology', cites Lewis Mumford's analysis of the transfer of spontaneity to the machine in *Technics and Civilization*.[125] Ellul draws on Mumford to argue that, while he correctly recognised the precarity of modernity and the influence of technology, his understanding of technology was too narrow.[126] And Arendt quotes recent writings by Mumford in *The New Yorker* in a late essay of hers. 'It is indeed only too true that the "premise underlying this whole age," its capitalist as well as its socialist development, has been "the doctrine of progress."'[127]

There are many other intellectual influences that connect these thinkers. References to Kierkegaard appear in Arendt's, Ellul's and Heidegger's work, while Adorno's first published work was *Kierkegaard: Construction of the Aesthetic*.[128] Heidegger and his students are notoriously nostalgic towards ancient Greek philosophy and society. Mumford's, Marcuse's and Adorno's work is imbued throughout with aesthetic analysis – in Mumford on architecture, for Adorno on music, while Heidegger and Arendt incorporate ideas of aesthetics into their work, such as in the observation that modern technology has diverged from its (ancient) conceptual origins in the Greek *technê*, craftsmanship or art. Feenberg also points out the similarities between certain aspects of Arendt's work – namely her late work drawing on Kant's aesthetic theory, and Marcuse's use of aesthetic categories. 'Is it a coincidence,' he asks:

> that both these Heidegger students affirm the disclosive power of art and attempt to transpose it to the political domain by drawing on Kant's third Critique? Where Arendt found a model of political judgment in Kant's theory of the imagination, Marcuse took the more radical route of applying that theory to technology.[129]

Among this group, then, a broad interweaving set of intellectual influences can be discerned that are of relevance to their analyses of technology. This, in some ways, laid the groundwork for their later critiques of technology and its influence on society and politics. But it was not until the events of the Second World War that the notion of technology as catastrophe would crystallise and emerge to its full extent in their writings.

A Moment of Transformation: The Impact of the Second World War

For most of our thinkers, the Second World War brought extraordinary upheaval, tragedy and hardship. This took different forms for each. Anders, Arendt, Jonas, Horkheimer, Adorno and Marcuse were forced by virtue of their Jewishness to leave Germany after the ascent of Hitler to power. While they took different paths – through Switzerland, France, England, Spain, Portugal, Canada, Palestine – staying in some countries for years at a time, all eventually ended up in the United States, most by the late 1930s, Jonas not until 1955.

Adorno, Horkheimer and Anders chose to return to Europe after the war, while Arendt, Jonas and Marcuse would remain, although they were officially stateless for many years until gaining their US citizenship. The Frankfurt School philosophers were fortunate, in relative terms. The Institute's funds gave them a degree of financial security and academic position which was afforded to few refugees. Arendt moved with no position to go to, no funds to draw on, and without even speaking English, which she learned only after her arrival in the United States. Yet she had managed to escape Europe with her immediate family (her husband and mother), and while life was far from comfortable, the trio managed adequately in New York. Anders also left Germany in 1933, to Paris, and onwards to the United States in 1936, where he worked variously as a labourer, tutor and writer, among other things, before he was employed by the New School in New York. Hans Jonas's journey was a little different, travelling via England to Palestine, fighting on the front in Italy and Germany during the war, and eventually emigrating to North America in the 1950s. While their fortunes varied, the devastating experience of exile was conveyed with pathos by Arendt in her 1943 essay 'We Refugees'.

> We lost our home, which means the familiarity of everyday life. We lost our occupation, which means the confidence that we are of some use in this world. We lost our language, which means the naturalness of reactions, the simplicity of gestures, the unaffected expression of feelings. We left our relatives in the Polish ghettos and our best friends have been killed in concentration camps, and that means the rupture of our private lives.[130]

Back in Europe, Ellul remained living in France. In 1934 and 1935, at the invitation of Protestant groups, Ellul visited Germany, even witnessing a Nazi rally on one occasion (as a critical observer).[131] In wartime France, he joined and served with the French Resistance, supporting and sheltering fleeing Jews.[132] His involvement was driven by his sense of Christian duty and not, he insisted, due to any political aims.[133]

Heidegger, as we know, took a very different path. Although his students later identified his latent authoritarianism in his early philosophy, his turn to Nazism came as a devastating shock to those who had earlier looked up to him. When, in 1933, Heidegger was elected as rector of Freiburg University by the Nazis, he carried out the requisite removal of Jewish professors from the institution including, notoriously, his former mentor and teacher, Edmund Husserl. His notebooks from the period reveal his profoundly antisemitic beliefs, although he would later attempt to distance himself from Nazi ideology. His leadership at Freiburg was unsuccessful, and he resigned a year after his election, although he remained both a member of the faculty and of the Nazi Party until the end of the war. 'In the early 1930s,' writes

Zimmerman, 'Heidegger believed that National Socialism offered an alternative to the technological nihilism forecast by Jünger,' and in the 1930s and beyond, Zimmerman adds, 'Heidegger interpreted Americanism and Bolshevism as different versions of the nihilism of modern technology – nihilism which could be overcome only by a complete transformation of German *Dasein*.'[134]

In America, Mumford was one step removed from the impact of the war. Nonetheless, the influences, connections and time that he had spent in Germany meant he felt its collapse deeply. Yet he was also severely critical of his own country's actions. America, too, had become uncivilised, he believed. 'By 1944,' writes Tschachler, 'and thus before Dresden and before Hiroshima and Nagasaki, Mumford interpreted the carpet-bombing of German cities by the U.S. Air Force as symptomatic of his country's decline into barbarism.'[135] The total war they were engaged in was itself totalitarian, he argued. Considering the fact that Mumford had once 'hailed Americans as the true heirs of the liberalism and humanitarianism of the Enlightenment, his about-turn is truly remarkable', Tschachler comments.[136] In 1944, when his son, Geddes, was killed on the Italian front at the age of nineteen, the tragedy only entrenched an already profound pessimism about the state of the world, including American politics.[137] Forman writes that, when Mumford read Ellul's work, 'he deplored [his] technological fatalism – he was silent about Heidegger's – . . . yet his position was essentially the same as theirs'.[138]

The events of the Second World War were, variously, personally, intellectually and politically world-shattering for these thinkers. It is clear their experiences were not the same: it would be absurd to suggest that the impact was comparable. But, for all these theorists, the war transformed how they thought about the world, what 'civilisation' really meant, and what the future might look like. The reality of total, global war, the Holocaust, the firebombing of cities, the dropping of the atomic bombs on Nagasaki and Hiroshima, effected the realisation among these individuals that anything – even the most unimaginably terrible – was possible. The catastrophe of the war and the rise of totalitarianism modelled a new future for these theorists. Even after German fascism was defeated, 'the specter of totalitarianism still hovered over the political landscape', materially in the Soviet Union, but also in the memories of those who had seen it rise to power with such apparent ease.[139] Technology played a major part in their concept of what that future might look like, and how in fact these possibilities had come to be. This shift reflected, in extremis, the fact that the war changed the whole context of thinking about technology and civilisation. The technological optimism of the Weimar era, specifically, was over. 'The break that shatters techno-romanticism was the conclusion of World War II and the new magnitude of impact by industrialized military technologies,' Don Ihde writes, '"Blitzkrieg," "The Battle of Britain," . . . the efficiencies of Holocaust gas chambers . . . the whole-city decimations of fire bombing and ultimately atomic bombing, all of which finally so numbed human sensitivity that techno-romanticism had to appear as not only antiquated but obscene.'[140]

COLD WAR CRITIQUES OF TECHNOLOGY

The critique of technology that this book explores is primarily located in the Cold War era. Yet, as this chapter has shown, the technology critique was not solely or initially a product of Cold War political thought. Technology as a force in politics was an important subject for political and social thought in Germany from at least the 1920s, and through the 1930s and 1940s, critiques of technology begin to be developed in the work of many of the thinkers centred here. The Second World War, and the rise of totalitarianism, further highlighted the threat technology posed to modern societies. But it was in the context of Cold War–era society and politics, with its explosion and spread of new technologies, combined with a narrative of a beneficent liberal-democratic political authority in the West, that this critique crystallised into something much more widespread, which spoke to the political issues of the era, albeit as a wider criticism of 'modernity'. The fear of total war merges seamlessly into the new context of an even greater threat, that of nuclear apocalypse. The many potential benefits of technology are either ignored in these narratives or reframed as threats. There is little recognition of the ways that technology might facilitate freedom, or improve the quality of life – as it undoubtedly did, in many ways, for billions of people in the twentieth century – or may act as a deterrent against war, as nuclear technology did. Generously, we might argue that the critics of technology believed the many benefits of technology to be obvious, well understood and not worth reprising – technological optimism ruled supreme, they believed. Less generously, omission of *any* discussion of the positive effects of technology on global humanity seems to leave the debate absurdly unbalanced, and this fact should certainly be noted.

Towards the end of the Second World War, and in the decades that followed, a wave of major publications concerning the threat of technology materialised. In 1944, Adorno and Horkheimer completed the first edition of what would later become *Dialectic of Enlightenment* (its final chapter, along with its title, was added in 1947). The year 1941, when the pair began to work on the manuscript together, was when, Jay argues, 'the Frankfurt School reluctantly acknowledged for the first time the crisis of the emphatic concept of reason that had been a mainstay of its work for much of the previous decade. The disillusionment was abrupt.'[141] The dilemma with which *Dialectic* was concerned was the 'self-destruction of the Enlightenment'.[142] Technology is central to this story. Modernity, Adorno and Horkheimer argue, has brought about the technological and material advance of progress, and an unprecedented degree of domination over nature. They point to the influence of the Baconian 'scientific attitude', in which 'knowledge, which is power, knows no obstacles . . . Technology is the essence of this knowledge', as well as to the demand for self-preservation implicit in Enlightenment reason, which, increasingly 'effected by the bourgeois division of labor . . . requires the self-alienation of the individuals who must model their body and soul according to the technical apparatus'.[143] The individual in contemporary society becomes nothing more than a facet of the technological capitalist society, and thus exists under the conditions of extreme domination.

In 1951, Mumford published *The Conduct of Life*, his first major post-war work dealing with technology, highlighting, like Adorno and Horkheimer, the consequences of the technological domination of nature. We have now 'reached a point in history where man has become his own most dangerous enemy', he argued, surrendering his higher capacities even as he boasts of conquering nature.[144] Only 'mass man' can emerge from such a society, he argues, 'incapable of choice, incapable of spontaneous, self-directed activities'.[145] He identifies a 'widespread ethical disintegration' occurring as a result of the creation of an 'interlocking machinery of schools, factories, newspapers and armies that have artificially destroyed the higher centers, have impaired the power of choice, have reduced symbolic functions to an almost reflex level, and have removed the capacity to co-ordinate from human to the machine process'.[146] This systematising and totalising automation is universal, Mumford argues, though masked by different ideologies: 'sometimes fascism, sometimes communism, sometimes capitalism or nationalism'.[147]

Similar themes are at the heart of Heidegger's analysis in 'The Question Concerning Technology', where he argues that modern technology must be understood as that which 'enframes' in a certain way, such that 'the work of technology reveals the real as standing-reserve'.[148] Technology imposes a particular order upon nature, he writes. For example: 'Agriculture is now the mechanized food industry. Air is now set upon to yield nitrogen, the earth to yield ore, ore to yield uranium, for example; uranium is set upon to yield atomic energy, which can be released either for destruction or for peaceful use.'[149] The problem with this is that it denies a more 'original revealing', and thus, '*precisely nowhere does man today any longer encounter himself, i.e., his essence*'.[150] Modernity is irrevocably intertwined with technology: 'everywhere we remain unfree and chained to technology, whether we passionately deny or affirm it.'[151]

In *The Technological Society*, Ellul argues (like Heidegger) that not technology in its particularity but rather technology as a '*totality of methods rationally arrived at and having absolute efficiency*' – which Ellul refers to as 'technique' – is what poses a threat to modern society.[152] Technique's singular principle is 'efficient ordering', leading to 'movement without direction'.[153] Thus, although the speed of life is accelerating, he writes, 'there is no orbit for it to take up, no point toward which it is heading, no place, no goal.'[154] Citing Mumford, Ellul insists that not capitalism but 'the machine' created our world and the 'most acute forms of human exploitation'.[155] Technology, as instantiated in modernity, has a dehumanising effect on humans; human nature is moulded to the machine, and humans are thereby alienated from themselves in an 'inhuman atmosphere'.[156]

After his return to Europe in 1950, Anders began to write his first major text, *The Obsolescence of Man*, published in 1956. Here, he offers a 'philosophical anthropology' of the condition of mankind in the atomic age. Contemporary technology, he suggests, has opened up a 'Promethean gap' between humanity's capacity to create and their ability to deal with the consequences of their creations. While the nuclear age's capacity for destruction is Anders's inspiration, the text deals with the effects of both productive and consumptive (media, leisure) technologies in modern society, and the ever-closer connection between the two, in the mode of technological domination.

Published two years later, Arendt's *The Human Condition*, the most systemic outline of her thinking on politics, was also written as an explicit response to a world in which radical technological changes – including nuclear power – gave the need to 'think what we are doing' a new urgency.[157] This rationale is important in understanding what influences lay behind *The Human Condition*. Yaqoob notes the influence of Heidegger on this work, but the context of American technological development, particularly automation, was also significant, and Brian Simbirski highlights the importance of Arendt's interest in 'automation and cybernetics'.[158] Since Galileo, she argues, when we first began to think of nature from a non-worldly perspective – the Archimedean point – we have become increasingly alienated from the earth.[159] This began with the telescope, but in the contemporary world Arendt points to the utilisation and probing of nuclear forces, manipulation of atomic particles, and understanding of the universe as ways in which science 'handle[s] nature from a point in the universe outside the earth'.[160] The atom bomb marks a further, but highly significant, development in modernity and earth alienation. While 'absolutely new in the history of science', it was also the culmination of a much longer trend.[161]

As the Cold War stretched on, the nuclear threat and its implications remained a major theme in the work of the critics of technology. However, both the influence of new technologies and that of changing cultural and political trends can also be seen. Arendt, for instance, had written in *The Human Condition*, that Sputnik, and man's first departure from the physical earth, was 'second in importance to no other' event in the modern era.[162] Her 1963 essay on 'The Conquest of Space' develops the ideas of earth alienation in the space age. 'We have come to our present capacity to "conquer space" through our new ability to handle nature from a point in the universe outside the earth.'[163] The issue with modern technology, namely that it attempts to produce a technological artifice modelled upon, or which unleashes, natural forces upon the human world, has shaped our entrance into space. As man becomes a universal creature, distanced ever further from his physical home, the earth alienation Arendt outlined in *The Human Condition* becomes increasingly tangible and dehumanising, she suggests.

In 1962, Ellul published *Propagandes*, his most extensive treatment of the propaganda system, which he associates with the technologies of contemporary media. The propaganda system, he writes, 'belongs to the technological universe, shares its characteristics, and is indissolubly linked to it'.[164] Technology makes propaganda possible, but more importantly and specifically propaganda fulfils the function of integrating the modern individual into a (technological) world in which they do not fit. 'Propaganda is a good deal less the political weapon of a regime (it is that also) than the effect of a technological society that embraces the entire man and tends to be a completely integrated society.'[165] It eases the sense of oppression of the technological system, he argues. As such, it is not something that is simply imposed upon populations: while mass society is a precondition for propaganda, in modernity there is also a 'sociological need for propaganda in the masses', because 'the individual left to himself is defenceless'.[166] Both America and the USSR are propaganda societies, he writes, although in different ways. 'The United States prefers to utilize the myth; the Soviet Union has for a long time preferred the reflex,' he writes, that is, to train the individual such that 'certain words, signs . . . persons or facts, provoke unfailing reactions'.[167]

Marcuse became the most famous of the Frankfurt School theorists when his work was taken up by the New Left in America and Europe in the 1960s. For a man reaching his seventh decade, this sudden surge of popularity among youthful radicals was rather surprising. But in the late 1960s, an undertone of optimism in Marcuse's thought, present in his analysis of the possibilities of technology, as well as his adherence to Marxist revolutionary ideals, surged. 'Marcuse's gloom about the demise of revolutionary opposition is dispelled,' writes Kellner, and his writings 'glow with revolutionary optimism'.[168] While he did not believe that the countercultural and political movements of the late 1960s were authentically revolutionary in themselves, Marcuse believed they might spark a new wave of radical and total upheaval of the kind required for a true revolution. This period marks the time of his greatest popularity, and his adoption as New Left 'guru' (despite his critiques of the movement as somewhat limited in ambition), and the peak of his revolutionary hopefulness. *One-Dimensional Man* was the most significant text of this period. In his analysis of the repression of dialectical – thus meaningful – existence in contemporary humanity, technology takes centre stage. In advanced industrial society, technology is not the sum-total of instruments, but rather 'a system which determines *a priori* the product of the apparatus, as well as the operations of servicing and extending it . . . [it] obliterates the opposition between the private and public existence'.[169] This makes it essentially 'totalitarian', he claims, and as such 'the traditional notion of the "neutrality" of technology can no longer be maintained . . . the technological system is a system of domination'.[170] Yet, a degree of hope remains, because political trends 'may be reversed', he writes; after all, 'essentially the power of the machine is only the stored-up and projected power of man. To the extent to which the work world is conceived of as a machine and mechanized accordingly, it becomes the *potential* basis of a new freedom for man.'[171]

Mumford's portrait of the modern 'megamachine' in his 1967 *The Myth of the Machine* characterised a technological society that was fundamentally repressive and dehumanising. Although this form of organisation dates from the ancient world, in modernity the megamachine is increasingly composed of mechanical parts, not human.[172] Egyptian pyramid building, for instance, 'was an archetypal example of simulated productivity', of which rocket building 'is our exact equivalent [today]'.[173] In these endeavours, individual freedom has no role.[174] While, like Arendt, Mumford dates the rise of the modern megamachine back to Galileo and the 'mechanical new world' of the sixteenth century that 'sought to understand, utilize and control the forces that derive ultimately from the cosmos and the solar system', its reassembly in practice took place from the eighteenth century onwards.[175] This is tracked through the French Revolution, the militarisation of society in the First World War and the emergence of totalitarianism in Germany and Russia in the 1930s – and, after the Second World War, in the West. The megamachine was rebuilt by the West on 'advanced scientific lines, with its defective human parts replaced by mechanical and electronic and chemical substitutes, and finally coupled to a source of power that made all previous modes of power production as obsolete as Bronze Age missiles'.[176] The 'Megatechnic complex' is today at the height of its power, he asserts. In its capacity for 'mass coercion and mass destruction . . . the system has nearly fulfilled its theoretical dimensions and possibilities; and if not judged by a more human measure, it is an overwhelming success'.[177] For Mumford, however, the progressive mechanisation of man, and the 'overcoming' of nature in fact represented the 'suicidal nihilism of our civilization'.[178]

Jonas's *The Imperative of Responsibility*, published in 1979, sought to construct a new ethic of responsibility that might function within the essentially changed context of a technological modernity. A new ethical imperative is required in this technological age, he writes, because, as technology empowers us, it also changes the character of human action.[179] Through lengthening and accelerating the speed and range of our actions, modern technology 'renders obsolete the tacit standpoint of all earlier ethics that, given the impossibility of long-term calculation, one should consider what is close at hand only and let the distant future take care of itself'.[180] In the face of a 'technical civilization that has become "almighty"', he writes, 'care for the future of mankind is the overriding duty of collective human action'.[181] From the 1960s onwards, Jonas developed a critique of the dangers inherent in technology itself: particularly its longer-term and unpredictable side effects.[182] Nuclear war might be catastrophic but was at least somewhat 'predictable' (in Jonas's words).

The profusion of technological critiques that emerged in the Cold War – of which I have highlighted only a few here – share a sense of what technology is, its key characteristics, and the claim that it is essentially at odds with both functional politics and free political action. By setting out the context and explaining some of the central claims of these authors in some of their works, I have tried to show the way in which their ideas of technology, while not identical, overlap in important ways, and together comprise an important mode of thinking about politics and society in the Cold War period. While these thinkers do not share broader fundamental premises in philosophy, politics or ethics, their understanding of technology and its relationship to modernity is core to their thinking and their understandings overlap and connect to one another in substantial ways. Feeding into this shared critique of technology are the various intellectual, personal and political influences outlined here: the intellectual influences they shared, particularly of Weimar Germany; the connections, friendships, relationships and antipathies that existed between the members of this group of intellectuals; and the dramatic political events of the 1930s and 1940s that demanded a response in their work, as well as their reaction against the West in the post-war and Cold War world, which identified in Western technology the heir to totalitarianism and its catastrophic politics. However, this was not a unitary endeavour, nor was it led by any one individual. In fact, to consider the technology critique in this way would be to diminish the complexity of thought on technology in these texts, but also to downplay the reality of the much more general influence or extent of this mode of thinking in social and political thought at this time, as a broadly significant intellectual trend and political critique.

NOTES

1. For example: Michael E. Zimmerman, *Heidegger's Confrontation with Modernity: Technology, Politics, Art* (Bloomington and Indianapolis: Indiana University Press, 1990); Don Ihde, *Heidegger's Technology: Postphenomenological Perspectives* (Bloomington: Indiana University Press, 1990); Jeffrey M. Shaw, *Illusions of Freedom: Thomas Merton and Jacques Ellul on Technology and the Human Condition* (Cambridge: The Lutterworth Press, 2014); Arthur M. Melzer, Jerry Weinberger and M. Richard Zinman (eds), *Technology in the Western Political Tradition* (Ithaca and London: Cornell University Press, 1993); Carl Mitcham, *Thinking Through Technology: The Path between Engineering and Philosophy* (Chicago: University of Chicago Press, 1994); Carl Mitcham and Robert Mackey (eds), *Philosophy and*

Technology: Readings in the Philosophical Problems of Technology (New York: The Free Press, 1972); Andrew Feenberg, *Technology, Modernity and Democracy* (London and New York: Rowman & Littlefield, 2018).

2. Jeffrey Herf, *Reactionary Modernism: Technology, Culture, and Politics in Weimar and the Third Reich* (Cambridge: Cambridge University Press, 1984).

3. As this work will show in the writings of, for instance, Arendt, Marcuse and Ellul.

4. Jan C. Schmidt, 'Ethics for the Technoscientific Age: On Hans Jonas' Argumentation and His Public Philosophy beyond Disciplinary Boundaries', in *Global Ethics and Moral Responsibility: Hans Jonas and His Critics*, ed. John-Stewart Gordon and Holger Burckhart (Abingdon: Routledge, 2013).

5. Waseem Yaqoob, 'The Archimedean Point: Science and Technology in the Thought of Hannah Arendt, 1951–1963', *Journal of European Studies* 44:3 (2014); Brian Simbirski, 'Cybernetic Muse: Hannah Arendt on Automation, 1951–1958', *Journal of the History of Ideas* 77:4 (2017).

6. For example: Christian Fuchs, *Critical Theory of Communication: New Readings of Lukács, Adorno, Marcuse, Honneth and Habermas in the Age of the Internet* (London: University of Westminster Press, 2016); Richard Wolin, *Heidegger's Children: Hannah Arendt, Karl Löwith, Hans Jonas, and Herbert Marcuse* (Princeton: Princeton University Press, 2001); Martin Jay, *The Dialectical Imagination: A History of the Frankfurt School and the Institute of Social Research, 1923–1950* (Berkeley: The University of California Press, 1973); Andrew Feenberg, *Heidegger and Marcuse: The Catastrophe and Redemption of History* (New York: Routledge, 2005); Langdon Winner, *Autonomous Technology: Technics-out-of-Control as a Theme in Political Thought* (Cambridge, MA: The MIT Press, 1977); Arthur M. Melzer, Jerry Weinberger and M. Richard Zinman (eds), *Technology in the Western Political Tradition* (Ithaca and London: Cornell University Press, 1993).

7. For example: Hannah Arendt, *The Promise of Politics* (New York: Schocken Books, 1993); Theodor Adorno, *The Jargon of Authenticity* (London: Routledge and Kegan Paul, 1973).

8. Jacques Ellul, *Anarchy and Christianity* (Grand Rapids: Eerdmans, 1992), 4.

9. Lewis Mumford, *Findings and Keepings: Analects for an Autobiography* (New York and London: Harcourt Brace Jovanovich, 1975), 315.

10. Ullrich Melle, 'Responsibility and the Crisis of Technological Civilization: A Husserlian Meditation on Hans Jonas', *Human Studies* 21:4 (1998), 343–4; Wolin, *Heidegger's Children*, 125.

11. Letter from Lewis Mumford to Hannah Arendt, 30 May 1965, Hannah Arendt Papers: Correspondence, 1938–1976; General, 1938–1976; 'Mo-Mu' miscellaneous, 1953–1975, undated. *The Hannah Arendt Archives* (Washington, DC: Library of Congress).

12. Samuel Matlock, 'Confronting the Technological Society', The *New Atlantis* 43 (Summer/Fall 2019).

13. Charles E. Silberman, 'Is Technology Taking Over?', *Educational Technology* 6:4 (1966), 8–9.

14. Jeffrey P. Greenman, Read Mercer Schuchardt and Noah J. Toly, *Understanding Jacques Ellul* (Eugene: Cascade Books, 2012).

15. Shaw, *Illusions of Freedom*, 21.

16. Greenman et al., *Understanding Jacques Ellul*, 19.
17. David Lovekin, *Technique, Discourse, and Consciousness: An Introduction to the Philosophy of Jacques Ellul* (Bethlehem: Lehigh University Press, 1991), 29.
18. Richard Polt, *Time and Trauma: Thinking through Heidegger in the Thirties* (London: Rowman & Littlefield, 2019), 89.
19. Ibid., 139.
20. Thomas S. W. Lewis, 'Mumford and the Academy', *Salmagundi* 49 (1980), 99.
21. Lewis Mumford, 'Introduction', in *Technics and Civilization* (San Diego, New York and London: Harcourt Brace Jovanovich, 1963), i.
22. Stephen Sheehan, 'The Nature of Technology: Changing Concepts of Technology in the Early Twentieth Century', *Icon* 11 (2005), 12.
23. Heinz Tschachler, '"Hitler Nevertheless Won the War": Lewis Mumford's Germany and American Idealism', *American Studies* 441:1 (1999), 100–1.
24. Ibid., 98.
25. Cited in Tschachler, 'Lewis Mumford's Germany and American Idealism', 98.
26. Tschachler, 'Lewis Mumford's Germany and American Idealism', 106.
27. Robert Casillo, 'Lewis Mumford and the Organicist Concept in Social Thought', *Journal of the History of Ideas* 53:1 (1992), 113.
28. Lovekin, *Technique, Discourse and Consciousness*, 128.
29. Ibid., 129.
30. Michael Morelli, *Theology, Ethics and Technology in the Work of Jacques Ellul and Paul Virilio: A Nascent Theological Tradition* (Lexington: Lexington Books, 2021), 32; 37.
31. Ibid., 38.
32. Ibid., 6–7.
33. Ibid., 7.
34. Ibid.
35. Jerry Weinberger, *Technology and the Problem of Liberal Democracy* (London: Routledge, 1988), 254.
36. Martin Heidegger, 'The Question Concerning Technology', in *The Question Concerning Technology and Other Essays*, trans. William Lovitt (New York: Harper & Row, 1977), 4.
37. Ibid., 12.
38. Andrew J. Mitchell, 'The Question Concerning the Machine: Heidegger's Technology Notebooks in the 1940s–1950s', in *Heidegger on Technology* (New York: Routledge, 2018).
39. Heidegger, 'The Question Concerning Technology', 16.
40. Lewis Mumford, *The Myth of the Machine: Technics and Human Development* (New York: Harcourt Brace Jovanovich, 1967), 188.
41. Lewis Mumford, *The Myth of the Machine: The Pentagon of Power* (New York: Harcourt Brace Jovanovich, 1970), 395–6.
42. Paul Forman, 'How Lewis Mumford Saw Science, and Art, and Himself', *Historical Studies in the Physical and Biological Sciences* 37:2 (2007), 272.
43. Jacques Ellul, *The Technological Society*, trans. John Wilkinson (New York: Random House, 1964), 3–4.
44. Ibid., xxv.

45. Jacques Ellul, *The Presence of the Kingdom*, trans. Olive Wyon (Philadelphia: Westminster Press, 1951), 41.
46. Ibid., 61.
47. Ibid., 13.
48. Jacques Ellul, *The New Demons*, trans. C. Edward Hopkin (New York: The Seabury Press, 1975), 66–7.
49. Ibid., 73.
50. Ibid.
51. John Abromeit, *Max Horkheimer and the Foundations of the Frankfurt School* (Cambridge: Cambridge University Press, 2011), 6.
52. Jay, *The Dialectical Imagination*, 217.
53. Ibid.
54. Caroline Ashcroft, 'From the German Revolution to the New Left: Revolution and Dissent in Arendt and Marcuse', *Modern Intellectual History* 19:3 (2022).
55. Wolf Schäfer, 'Stranded at the Crossroads of Dehumanization: John Desmond Bernal and Max Horkheimer', in *On Max Horkheimer*, ed. John McCole, Seyla Benhabib and Wolfgang Bonß (Cambridge, MA: MIT Press, 1993), 167.
56. Ibid.
57. Theodor W. Adorno and Max Horkheimer, *Dialectic of Enlightenment* [1944/47] (London: Verso, 1997), 4.
58. Max Horkheimer, *Eclipse of Reason* (London: Continuum, 1947), v.
59. Adorno and Horkheimer, *Dialectic of Enlightenment*, 30.
60. Ibid., 121–2.
61. Ibid., 123.
62. Ibid., 126.
63. Theodor Adorno, *Prisms* (Cambridge, MA: MIT Press, 1981), 34.
64. Theodor Adorno, *Negative Dialectics* (London: Routledge, 1973), 362.
65. Adorno and Horkheimer, *Dialectic of Enlightenment*, 121.
66. Richard Wolin, *The Frankfurt School Revisited, and Other Essays on Politics and Society* (London and New York: Routledge, 2006), 83.
67. Herbert Marcuse, 'Some Social Implications of Modern Technology', *Studies in Philosophy and Social* Science 9 (1941), 418.
68. Ibid., 422
69. Herbert Marcuse, *One-Dimensional Man* (Boston, MA: Beacon, 1964), xvi.
70. Wolin, *Heidegger's Children*, 13.
71. Jason Dawsey, 'After Hiroshima: Günther Anders and the History of Anti-Nuclear Critique', in *Understanding the Imaginary War: Culture, Thought and Nuclear Conflict, 1945–90*, ed. Matthew Grant and Benjamin Ziemann (Manchester: Manchester University Press, 2016), 141.
72. John-Stewart Gordon, 'Introduction', in *Global Ethics and Moral Responsibility*, 22.
73. Ibid., 5
74. Jason Dawsey, 'Ontology and Ideology: Günther Anders's Philosophical and Political Confrontation with Heidegger', *Critical Historical Studies* 4:1 (Spring 2017), 8.
75. Christopher John Müller, 'Introduction', in *Prometheanism: Technology, Digital Culture and Human Obsolescence* (London and New York: Rowman & Littlefield, 2016), 7.

76. Dawsey, 'After Hiroshima', 42.
77. Hans Jonas, 'Introduction', in *Philosophical Essays* (New York and Dresden: Atropos Press, 1980), xiii.
78. Hans Jonas, 'Prologue', in *Mortality and Morality: A Search for Good after Auschwitz*, ed. Lawrence Vogel (Evanston: Northwestern University Press, 1996), 49–50.
79. Ibid., 46.
80. Hans Jonas, 'Technology and Responsibility: Reflections on the New Tasks of Ethics', in *Philosophical Essays* (New York and Dresden: Atropos Press, 1980), 8.
81. Ibid., 9.
82. Ibid.
83. Jonas, 'Seventeenth Century and After', 75.
84. Hans Jonas, 'Socio-Economic Knowledge and Ignorance of Goals', in *Philosophical Essays* (New York, Dresden: Atropos Press, 1980), 106.
85. Hans Jonas, *The Imperative of Responsibility* (Chicago: University of Chicago Press, 1984), 1.
86. Ibid., 17.
87. Ibid., 201.
88. Dawsey, 'After Hiroshima', 142.
89. Ibid., 153.
90. Günther Anders, *Die Antiquiertheit des Menschen*, vol. 2: *Über die Zerstörung des Lebens im Zeitalter der dritten industriellen Revolution* [1980] (Munich: C. H. Beck, 2018), 279–80
91. Herf, *Reactionary Modernism*.
92. Tschachler, 'Lewis Mumford's Germany and American Idealism', 100–1.
93. Lewis Mumford, *The Conduct of Life* (New York: Harcourt, Brace & Co., 1951), 149.
94. Tony Judt, *Past Imperfect: French Intellectuals, 1944–1956* (Berkeley: University of California Press, 1992), 196.
95. Ellul also drew widely on contemporary French thinkers; Jean Fourastie and Georges Friedmann, for instance, feature in *The Technological Society*.
96. Lovekin, *Technique, Discourse and Consciousness*, 90.
97. Herf, *Reactionary Modernism*, 1–2.
98. Ibid., 44.
99. Ibid., 38.
100. Zimmerman, *Heidegger's Confrontation with Modernity*, 80.
101. Ibid., 216–17.
102. Mikko Immanen, *Toward a Concrete Philosophy: Heidegger and the Emergence of the Frankfurt School* (Ithaca: Cornell University Press, 2020), 3.
103. Wolin, *The Frankfurt School Revisited*, 5; Adorno, *The Jargon of Authenticity*, xix.
104. Wolin, *The Frankfurt School Revisited*, 12.
105. Ibid., 79.
106. Hannah Arendt, 'What is Existential Philosophy?' [1946], in *Essays in Understanding 1930–1954: Formation, Exile, and Totalitarianism* (New York: Schocken Books, 2005), 181.

107. Yaqoob, 'The Archimedean Point', 205.
108. Jay, *The Dialectical Imagination*, 25; Jay also points out here that this was explicitly in opposition to Heidegger's 'yearning for a return to the comfort of meaningful unities'.
109. Martin Jay, *Reason after Its Eclipse: On Late Critical Theory* (Madison: University of Wisconsin Press, 2016), 97.
110. Abromeit, *Max Horkheimer and the Foundations of the Frankfurt School*, 401.
111. Jay, *The Dialectical Imagination*, 193.
112. Ibid.
113. Barry Katz, *Herbert Marcuse: Art of Liberation* (London: Verso, 1982), 217–18.
114. Herbert Marcuse, 'Socialist Humanism', in *Socialist Humanism: An International Symposium*, ed. Erich Fromm (Garden City: Doubleday, 1965), 112.
115. Ibid.
116. Ibid., 116.
117. Ibid., 115.
118. Ibid., 116.
119. Lovekin, *Technique, Discourse and Consciousness*, 127.
120. Greenman et al., *Understanding Jacques Ellul*, 83.
121. James Muldoon, 'The Origins of Hannah Arendt's Council System', *History of Political Thought* 37:4 (2016); Ashcroft, 'Revolution and Dissent in Arendt and Marcuse'.
122. Hannah Arendt, *On Revolution* (New York: Pelican Books, 1973), 65.
123. Hans Jonas, 'Reflections on Technology, Progress and Utopia', *Social Research* 48:3 (1981), 450.
124. This was due to illness: Mumford caught tuberculosis during his university studies at the New School for Social Research in New York.
125. Marcuse, 'Some Social Implications of Modern Technology', 417.
126. Ellul, *The Technological Society*, 42
127. Hannah Arendt, 'Home to Roost' [1975], in *Responsibility and Judgment* (New York: Schocken Books, 2005), 262–3.
128. Theodor Adorno, *Kierkegaard: Construction of the Aesthetic* [1933] (Minneapolis: University of Minnesota Press, 1989); Hannah Arendt, 'What is Existential philosophy?' [1946], in *Essays in Understanding*, 169; Jacques Ellul, *The Technological Bluff*, trans. Geoffrey Bromiley (Grand Rapids: Eerdmans, 1990), 411–12.
129. Feenberg, *Heidegger and Marcuse*, 96.
130. Hannah Arendt, 'We Refugees', in *The Jewish Writings* (New York: Schocken Books, 2007), 264–5.
131. Greenman et al., *Understanding Jacques Ellul*, 85–6.
132. Lovekin, *Technique, Discourse and Consciousness*, 135.
133. Ibid.
134. Zimmerman, *Heidegger's Confrontation with Modernity*, 36; 42.
135. Tschachler, 'Lewis Mumford's Germany and American Idealism', 98.
136. Ibid.
137. Forman, 'How Lewis Mumford Saw Science, and Art, and Himself', 311.

138. Ibid.
139. Stephen E. Bronner, *Reclaiming the Enlightenment: Toward a Politics of Radical Engagement* (New York, Columbia University Press, 2006), 95.
140. Don Ihde, *Heidegger's Technologies: Postphenomenological Perspectives* (New York: Fordham University Press, 2020), 15.
141. Jay, *Reason after Its Eclipse*, 97.
142. Adorno and Horkheimer, *Dialectic of Enlightenment*, xiii.
143. Ibid., 4; 30.
144. Mumford, *The Conduct of Life* (New York: Harcourt, Brace & Co., 1951), 11.
145. Ibid., 16.
146. Ibid., 144.
147. Ibid., 149.
148. Heidegger, 'The Question Concerning Technology', 21.
149. Ibid., 14–15.
150. Ibid., 27.
151. Ibid., 4.
152. Ellul, *The Technological Society*, 3–4.
153. Ibid., 110.
154. Ibid., 5.
155. Ibid.
156. Ibid., 4.
157. Hannah Arendt, *The Human Condition* (Chicago: University of Chicago Press, 1998), 5.
158. Yaqoob, 'The Archimedean Point', 217.
159. Arendt, *The Human Condition*, 257–68.
160. Arendt, 'The Conquest of Space and the Stature of Man', in *Between Past and Future* (New York: Penguin, 2006), 273.
161. Hannah Arendt, 'Introduction *into* Politics', in *The Promise of Politics*, 154.
162. Arendt, *The Human Condition*, 1.
163. Arendt, 'The Conquest of Space and the Stature of Man', 273.
164. Jacques Ellul, *Propaganda: The Formation of Men's Attitudes*, trans. Konrad Kellen and Jean Lerner (New York: Knopf, 1965), x.
165. Ibid., xvii.
166. Ibid., 121; 96; 92.
167. Ibid., 31–2.
168. Douglas Kellner, *Herbert Marcuse and the Crisis of Marxism* (London: MacMillan, 1984), 285. The specific essays Kellner refers to here are 'Five Lectures' and 'An Essay on Liberation'.
169. Marcuse, *One-Dimensional Man*, xv.
170. Ibid., xvi.
171. Ibid., 2.
172. Lewis Mumford, *The Myth of the Machine: Technics and Human Development* (New York: Harcourt Brace Jovanovich, 1967), 188.
173. Ibid., 205.
174. Ibid., 183.

175. Lewis Mumford, *The Myth of the Machine: The Pentagon of Power* (New York: Harcourt Brace Jovanovich, 1970), 39; 4.
176. Ibid., 245.
177. Ibid., 346.
178. Lewis Mumford, *My Works and Days: A Personal Chronicle* (New York and London: Harcourt Brace Jovanovich, 1979), 16.
179. Jonas, *The Imperative of Responsibility*, 1.
180. Ibid., 34.
181. Ibid., 136.
182. Jonas, cited in Carl Mitcham, 'Foreword', in Hans Jonas, *Philosophical Essays* (New York and Dresden: Atropos Press, 1980), xvii.

2

HISTORICAL NARRATIVES OF TECHNOLOGICAL DEVELOPMENT

Introduction

Although the critique of technology is rooted firmly in a twentieth-century context, modern technology and its influence on politics and society are claimed to have a much longer history by these theorists. This chapter will explore how the critics of technology position technology in modernity and its role in the emergence of modernity by outlining their histories of the development of technology. These historical narratives, although similar in many ways, also diverge in some respects – different authors, for example, prioritise analysis of certain historical periods over others, or particular trends or concepts. The narratives they offer conflict with one another in some ways – but not, this chapter will suggest, in ways that undermine the substantial agreement between how they perceive technology's influence on modernity. The way in which modernity breaks with that which comes before is integrally connected to the development of technology, they argue.

None of the thinkers included in this study are primarily historians; yet all of them would consider historical understanding to be central to their work. For Heidegger and Arendt, recourse to ancient Greek and Roman history underpins their philosophy and politics; Jonas's early work on Gnosticism inspired his later work on technology; Adorno and particularly Horkheimer's understanding of eighteenth-century history was foundational to their thinking on technology, modernity and reason; Mumford's understanding of ancient history was the starting point for his thinking on the modern megamachine;

63

Ellul's *The Technological Society* traces technology through the social and economic changes of the modern period; while even Marcuse relied upon historical analysis in his writings on the development of contemporary civilisation.[1] These histories, it should be said, have often been considered questionable. It has often been observed that the philosophy or politics of ancient Greece was radically idealised by some of these thinkers, for instance, or that their analysis of events such as the French Revolution is highly partisan. It is a highly, even exclusively, Westernised account of the development of technology, which has its own limitations. And it prioritises an account of novelty or innovation in technology which, as David Edgerton has persuasively argued, can present a distorted understanding of the reality of technology's character and influence in any particular period of time.[2] With this in mind, this chapter is not a critique or a defence of these histories, but merely an outline of the way in which these theorists imagine technology to have been realised in modern history, presented in order to show how they envisage the development of modernity to be interwoven with the development of technology. This historical framing of technology as a determinant of the modern form of politics is at the heart of these thinkers' critiques of technology.

The Prehistory of Modern Technology

The ancient world and technê

Although the term 'technology' is a modern term, in the sense our critics of technology are using it, its etymology can be traced back to the Greek *technê*. *Technê* refers to craftsmanship or the practical arts; it is a form of practical knowledge. This is distinct from *epistêmê*, on the one hand, knowledge of necessary or scientific truth, and *poiesis*, or artistic creation, on the other. However, the precise meaning of *technê* and the relationship between these terms varies between Greek philosophers. *Technê*, in the historical reading of the critics of technology, can appear either as a mode of tool-usage quite at home in the human world or, alternatively, as something that contains dangers in and of itself.

Heidegger's interpretation of *technê* is foundational to his analysis of technology in his writings of the 1940s and 1950s. 'Technology [*Technik*] understood as modern, i.e., as the technology of power machines . . . [refers] back to a hidden essence of technology that encompasses what the Greeks already called *technê*,' he writes.[3]

> *Technê* is the name not only for the activities and skills of the craftsman, but also for the arts of the mind and the fine arts. *Technê* belongs to bringing-forth, to *poiesis*; it is something poietic . . . what is decisive in *technê* does not lie at all in making and manipulating nor in the using of means, but rather in the aforementioned revealing. It is as revealing, and not as manufacturing, that *technê* is a bringing-forth.[4]

Heidegger thus makes a claim that modern technology, just as ancient *technê*, is essentially a 'revealing'. Yet, he writes, while the early Greeks lived 'immediately in the openness of phenomena', towards the end of the classical period an instrumentalist *technê* began to make its way into the new metaphysical understanding of the philosophers, and particularly Plato.[5] Greek metaphysics was thus viewed by Heidegger as 'proto-technological', Zimmerman explains. For Heidegger, 'the history of metaphysics became the history of the unfolding of *productionist* metaphysics'.[6] In Greek philosophy, *making* began the long journey which would conclude with its supplanting *being*: the progressive decline of humanity's understanding of what it is 'to be'. In this way, the source of modern technology, Heidegger believes, can be identified in ancient Greece, although, of course, it was not realised as what we would recognise as technology until the contemporary era.

Adorno and Horkheimer also argue that the origin of the shift away from a primarily magical or mythological frame of reference, towards the rationalised model of technological modernity, began long before modern technology itself. Yet in contrast to Heidegger, for Adorno and Horkheimer, 'despite their deep pessimism about the future, the past could not offer a refuge from the present', Gerard Delanty and Neal Harris argue. 'There is no indication of nostalgia for a pre-technological past in the Frankfurt School texts,' they write.[7]

Arendt and Jonas, though followers of Heidegger, offer a more prosaic reading of *technê*, closer to the contemporary idea of technology. The difference between then and now, they both agree, was that the Greeks recognised the limitations or dangers of the activity. The Greeks rejected *technê*, argues Arendt, because they considered instrumentalism to be *banausic*, a philistine activity; even 'the great masters of Greek sculpture and architecture were by no means excepted from the verdict'.[8] In the Greek world, writes Jonas, '*technê* was a measured tribute to necessity, not the road to mankind's chosen goal . . . Now, *technê* in the form of modern technology has turned into an infinite forward-thrust of the race.'[9] He contrasts the Greek *polis*, 'this supreme work of collective "art" wrested from nature in the first flowering of Western man', with the 'extreme artificiality of our technologically constituted, electronically integrated environment'.[10]

Ellul's history of technology takes a different starting point. 'If we go back to the beginning, "technique,"' writes Ellul, 'is essentially Oriental: it was principally in the Near East that technique first developed and it had very little in the way of scientific foundation. It was entirely directed toward practical application and was not concerned with general theories.'[11] The Greeks, he argues, although possessing a coherent idea of science, separated science, a contemplative activity, from technique.[12] The Greeks were suspicious of technology, because of its characteristic of 'brute force [that] . . . implied a want of moderation', in a society where self-control was deemed the supreme virtue, he argues.[13] 'The rejection of technique was a deliberate, positive activity involving self-mastery, recognition of destiny, and the application of a given conception of life.'[14] In Rome, however, technique takes on a very different hue for Ellul. Whereas previously social technique was limited, 'in Rome . . . we pass on, at one step, to the perfection of social technique, both civil and military. Everything in Roman society was related to Roman law in its multiple forms, both public and private.'[15] They ruled by organisation, technique, rather than force, because it was more 'economical' to do so, he argues.[16]

For Heidegger, Adorno and Horkheimer, technology was something 'un-Greek', opposed to the authentic nature of Greek society. For Arendt, Jonas, and Ellul, the same could be said to be true, with the important distinction that they believed the Greeks *recognised* the non-political nature of technology, although Ellul locates the rise of technique in the Roman world instead. Mumford offers a different story. He also locates the origins of the modern technological in the ancient world. But *his* chronology of technology does not track back to Greek metaphysics or Roman social organisation, nor does he harbour nostalgia for a pre-technological ancient past. 'What economists lately termed the Machine Age or the Power Age', Mumford claimed, 'had its origin, not in the so-called Industrial Revolution of the eighteenth century, but at the very outset in the organization of an archetypal machine composed of human parts': the 'megamachine'.[17] Both ancient and modern megamachine are based upon production *and* destruction, he writes. 'While the labor machine largely accounts for the rise of "civilization," its counterpart, the military machine, was mainly responsible for the repeating cycles of extermination, destruction, and self-extinction.'[18] These are the 'poles' of civilisation, then, both 'mechanical' in nature. The difference between the ancient and modern megamachine was that the ancient megamachines were organisations of men, whereas modern megamachines are constructed (partly) of 'the mechanization of their working instruments', *alongside* organisation of men.[19] In both instances, the megamachine is essentially an authoritarian, even totalitarian, form of organisation: 'Individual initiative and responsibility had no place . . . The dedicated members of the megamachine, early and late, remain Eichmanns: doubly degraded because they have no consciousness of their own degradation.'[20]

Technê, then, always appeared to harbour dangers for society, albeit these dangers were recognised and thus mitigated, *or* unrecognised but less significant in the underdeveloped *technê* of the classical world. Even in the starkly different and much darker story that Mumford offers, while the ancient 'megamachine' is an authoritarian and dehumanising structure, it has not reached the excesses of mechanical efficiency that the modern megamachine has, with regards to either destruction or production.

Christianity: worldliness and unworldliness

It is notable that processes of secularisation are seen to have a significant role in the development of technologised society in the thought of all of these theorists. Yet, their claims about how the rise and dominance of Christianity in the medieval period influenced technological development disagree completely. The interaction between Christianity and technology is seen to operate in completely different ways, and understanding this is relevant to interpreting their later work on the influence of secularisation – some thinkers view the development of Christianity as integral to the continued development of technique; others perceive it as causing the reversal of earlier trends to technologisation.

Ellul argues that after the fall of Rome (where technique had flourished), and the rise of Christianity, technique faced a reversal of its fate. There was a 'complete obliteration' of technique because 'Christians held judicial and other technical activity in such contempt that they were considered the "enemies of the human race"'. As such, society from the fourth to the fourteenth century was characterised by 'a total absence of the technological will. It was "a-capitalistic" as well as "a-technical."'[21] Technical weakness, although lessening, persisted all the way into the beginning of the eighteenth century.[22] Heidegger, in contrast, believes that the Greek proto-productionist metaphysic only grew in strength, first gaining extraordinary impetus in the Roman world, where metaphysics was increasingly understood in terms of cause and effect, and later in the Christian world. Zimmerman argues that for Heidegger:

> Christian theologians, including St. Augustine and St. Thomas Aquinas, furthered this Latin representation of being as the productive basis for things. For such theologians, God became identified with the being of entities, i.e. the self-grounding Creator who produced all creatures . . . God was reduced to the status of an all-powerful causal agent.[23]

Arendt, on the other hand, like Ellul did not hold Christianity to be an instrumentalist ideology, but did argue that Christianity's absolutisation of the sacredness of life would remain a fundamental premise even in secular modernity. 'Life' has become, and remains, 'the highest good of modern society', she claims, because the Christian 'fundamental belief in the sacredness of life has survived, and even remained completely unshaken by, secularization and the general decline of Christian faith'.[24] The ascendency of 'life itself', or the life process, had a transformational effect on politics and the nature of technology, she would argue; the influence of Christianity on technology survived the decline of the religion itself. For Mumford, too, the influence of Christianity was double-edged. While Christianity remained a dominant belief, the megamachine was restrained. 'The medieval social order could not be completely mechanized or depersonalized because it was based, fundamentally, upon a recognition of the ultimate value and reality of the individual soul,' he argued.[25] Yet Christianity would also add to the original model of the megamachine 'the one element that was lacking: a commitment to moral values and social purposes that transcended the established forms of civilization'.[26] By supposedly renouncing power, Mumford argues, 'it augmented power in the form that could be more widely distributed and more effectively controlled in machines', making the machine at the same time more acceptable.[27] Like Arendt, though, Mumford believes that this would become most significant in the wake of Christianity's decline. In the 'bare, depopulated world of matter and motion', he writes, 'mechanical invention, even more than science, was the answer to a dwindling faith and a faltering life-impulse'.[28]

THE SCIENTIFIC REVOLUTION AND 'GALILEO'S CRIME'

In the development of modern technology, the scientific revolution of the seventeenth century is another critical juncture; modern science and technology are interlocked from this period. But it is emphatically *not* the case that the emergence of science precedes, or lays the groundwork for, the later development of technology, as per the 'primacy of science' model that Paul Forman has argued was so characteristic of the modern era. Rather, perhaps pre-empting a later postmodern shift to the cultural primacy of technology that Forman argues took place around 1980, this group claim the reverse is true: technology is the basis of modern science.[29] Their claims, moreover, are not without ground, as Feenberg points out: 'historians have shown that few technologies arose as applications of science until quite recently . . . [and] science is even more dependent on technology today than in the past.'[30]

For most of the critics of technology, Galileo Galilei is a key figure, along with Francis Bacon. In German philosophy, the fascination with Galileo, and the claim that technologisation accompanies the new science, precedes much of the work of the critics of technology on the subject. In his 1936 *The Crisis of European Sciences and Transcendental Phenomenology*, Husserl writes that, with Galileo, 'a technization takes over all other methods belonging to natural science. It is not only that these methods are later "mechanized." To the essence of all method belongs the tendency to superficialize itself in accordance with technization.'[31] Ernst Cassirer, too, wrote in 1930 that:

> Galileo . . . began from technical problems . . . one must keep in mind the fact that each of Galileo's discoveries in physics and astronomy are closely linked to some instrument of his own invention or to some special set-up. His technological genius is the authentic prerequisite for the scientific efforts through which his theoretical originality first received its direction and expression.[32]

In *Dialectic of Enlightenment*, Adorno and Horkheimer cite Husserl: 'In Galileo's mathematization of nature, nature itself is idealized on the model of the new mathematics.' Thus, they continue, 'thought is reified as an autonomous, automatic process, aping the machine it has itself produced, so that it can finally be replaced by the machine.'[33] But in the *Dialectic* the personification of the scientific revolution is Bacon. For Bacon, the human mind is 'to hold sway over a disenchanted nature'.[34] From the first, science and technology are intimately associated: 'Technology is the essence of this knowledge,' Horkheimer asserts. 'It does not work by concepts and images, by the fortunate insight, but refers to method, the exploitation of others' work, and capital ... What men want to learn from nature is how to use it in order wholly to dominate it and other men.'[35] Despite science's claim to authoritative validity, it in fact represses truth, in its seeking out of 'universal interchangeability', whereby 'the multitudinous affinities between existents are suppressed by the single relation between the subject who bestows meaning and the meaningless object, between rational significance and the chance vehicle of significance'.[36] As Horkheimer writes, 'Thought becomes illusionary whenever it seeks to deny the divisive function, distancing, and objectification.'[37] Mathematics was the model of this kind of thinking ('essential to the expansion of industry', he adds), but once it becomes characteristic of thinking itself, or the only form of thought, 'it takes on a kind of materiality and blindness, becomes a fetish, a magic entity that is accepted rather than intellectually experienced'.[38] Science and reason's purpose is now – and since Bacon's time, has always been – to dominate nature. Within this, Horkheimer argues, although man seeks to distance or differentiate himself from nature, is included man: 'the history of man's efforts to subjugate nature is also the history of man's subjugation by man.'[39] Today, the subject has become 'objectified' into 'technical process':[40]

> Today, when Bacon's utopian vision that we should 'command nature by action' – that is, in practice – has been realized on a tellurian scale, the nature of the thralldom that he ascribed to unsubjected nature is clear. It was domination itself. And knowledge, in which Bacon was certain the 'sovereignty of man lieth hid,' can now become the dissolution of domination. But in the face of such a possibility, and in the service of the present age, enlightenment becomes wholesale deception of the masses.[41]

Arendt, well versed in both modern physics and the history of science, points to Galileo's advances as a transformational moment for modernity. 'Modern natural science owes its great triumphs to having looked upon and treated earthbound nature from a truly universal viewpoint, that is, from an Archimedean standpoint taken, wilfully and explicitly, outside the earth,' she writes.[42] It was from this starting point that what Arendt refers to as 'earth-alienation' came to dominate modern science, and by which 'every science, not only physical and natural science, so radically changed its innermost content that one may doubt whether prior to the modern age anything like science existed at all'.[43] It was Galileo's telescope that first enabled us to step outside the world, to know with all the certainty of sense perception the wider universe beyond our own earth. At the same time, science developed the experimental method, by which it began to 'prescribe conditions and to provoke natural processes'. This would ultimately lead into, centuries later, the unleashing of atomic energy. 'What then developed into an ever-increasing skill in unchaining elemental processes . . . has finally ended in a veritable art of "making" nature, that is, of creating "natural" processes which without men would never exist and which earthly nature by herself seems incapable of accomplishing.'[44] For Arendt, then, modern science was always bound up with technology; science does not precede technology. The scientific revolution emerged from technological invention and development – the telescope, experimental apparatus – rather than theoretical models. Of the great changes that took place at the dawn of the modern period, we find in retrospect that:

> these first tentative glances into the universe through an instrument, at once adjusted to human senses and destined to uncover what definitely and forever must lie beyond them, set the stage for an entirely new world and determined the course of other events, which with much greater stir were to usher in the modern age.[45]

While Heidegger believed that the development of technology can be traced back to Greece, with the emergence of science in the early modern era, he notes that the technological impulse took another great leap forward. 'With the exception of the beginning among the Greeks, the rise of modern science became decisive for the essential definition of the thing.'[46] According to the new scientific view, 'the thing is material, a point of mass in motion in the pure space-time order, or an appropriate combination of such points. The thing so defined is then considered as the ground and basis of all things.'[47] And because the scientific project 'established a uniformity of all bodies according to relations of space, time, and motion, it also makes possible and requires a universal uniform measure as an essential determinant of things, i.e. numerical measurement'.[48] In his reflections on mathematics, Heidegger concludes, 'Descartes grasps the idea of a *scientia universalis*, to which everything must be directed and ordered as the one authoritative science.'[49] Thus, science begins to take on the qualities of technology that Heidegger identifies: its 'enframing' of the world. While 'machine-power technology' does not develop until the eighteenth century, he writes, 'modern technology, which for chronological reckoning is the later, is, from the point of view of the essence holding sway within it, the historically earlier'.[50] While Descartes, rather than Galileo or Bacon, appears as the pivotal figure here, it is worth noting that Heidegger draws from Cartesianism (in the context of the development of technology) very similar conclusions to those that Arendt, Adorno and Horkheimer drew from the influence of Galileo and Bacon. This is notable, given the distinctiveness of the traditions of empiricism and rationalism that they inspired, but also highlights the continuity of the historical narrative that the critics of technology adhere to. Different characters play pivotal roles, but the roles are – in this particular sense – the same. For Heidegger, whose attempt to rediscover being in its unified, foundational form was oriented around a critique of Cartesian dualism, it is perhaps not surprising that Descartes should appear as a central figure in this history as well.

Jonas appears to describe a somewhat different historical relationship between the development of modern science and modern technology. The rise of science historically preceded the rise of (modern) technology, he writes, and it was not until the nineteenth century that we see 'the breakthrough of mature science into technology, and thereby the rise of modern science-infused technology itself'.[51] The scientific revolution 'changed man's ways of thinking, *by* thinking, before it materially changed, even affected, his ways of living,' he observes. 'It was a change in theory, in world-view.'[52] It is in fact a 'misconception', he argues, that the 'evolutions of modern technology went hand in hand'. Rather, the scientific breakthroughs took place in the seventeenth century, and 'the breakthrough of mature science into technology' in the nineteenth century.[53] 'Technology, historically speaking, is the delayed effect of the scientific and metaphysical revolution with which the modern age begins,' he writes.[54] And while the scientific revolution began 'in acts of supreme and daring freedom', it has now 'set up its own necessity and proceeds on its course like a second, determinate nature – no less deterministic for being man-made'.[55] Crucially, Jonas believes, technology has changed how we understand and relate to nature. From modern science, a new concept of nature emerged, which 'contained manipulability at its theoretical core and, in the form of experiment, involved actual manipulation in the investigative process'.[56] Nature was no longer simply the environment in which we lived, or the bodies in which we lived, it was not a given, but rather a resource to utilise.

Yet, while Jonas distinguishes the development of science and technology historically, he also is at pains to show that the development of technology, after the scientific revolution, was inevitable. The development of modern technology in the nineteenth century 'was far from accidental, or extraneous to the cause. The technological turn was somehow in the cards from the beginning,' he insists.[57] Technology was 'implied as a possibility in the metaphysics, and trained as a *practice* in the procedures, of modern science'.[58] Modern science and modern technology cannot therefore in truth be separated. 'To modern theory in general, practical use is no accident but is integral to it,' he writes; in other words, '"science" is technological by its nature.'[59]

Ellul, Mumford and Marcuse have greater faith in the potential of science. As Fred Alford writes, 'for Marcuse the basic structure of science is historically relative. A revolutionary change in social relations could bring with it a revolutionary new science as well.'[60] Ellul identifies the origins of modern science in the harmless pursuit of philosophical understanding.[61] And Mumford, insists Paul Forman, 'no matter how severely he chastised science and scientists, was fundamentally on their side'.[62] Yet they all argued that the character of modern science was essentially technological and a transformative influence on the contemporary world. Science played a central part in the development of the modern megamachine, Mumford wrote, in its production of 'abstract symbols, rational systems, universal laws, repeatable and predictable events, [and] objective mathematical measurements', which he traces back to Galileo.[63] Even though Galileo had 'no anticipation of what the world would be like if his standards were universally accepted . . . he had nevertheless, in dismissing subjectivity excommunicated history's central subject, multidimensional man'.[64] The resulting machine 'succeeded in de-naturing or banishing every organic attribute . . . [resulting in] an environment like our present one: fit only for machines to live in'.[65] Ellul in is agreement. Although the scientific method was first applied in order to prove certain philosophical hypotheses, it was not long before it was applied for 'usefulness', he writes.[66] 'From then on science was lost!' he claims.[67] It became increasingly about effectiveness itself. 'Technique began to develop and extend itself only after science appeared,' he wrote; 'to progress, technique had to wait for science.'[68] Yet, he also writes that technique, in practical terms, *precedes* science, a change that takes place at the point at which science becomes experimental.[69]

> In every instance it is clear that the border between technical activity and scientific activity is not at all sharply defined. When we speak of technique in historical science, we mean a certain kind of preparatory work: textual research, reading, collation, study of monuments, criticism and exegesis. These represent an ensemble of technical operations which aim first at interpretation and then at historical synthesis, the true work of science. Here, again, technique comes first . . . Where the technical means do not exist, science does not advance.[70]

While, for Marcuse, science can be considered benign in some cases, its historical development in modernity is not. The development of contemporary technology was presaged by other changes brought about by *seemingly* more benign forms of scientific rationality. 'Long before technological man and technological nature emerged as the objects of rational control and calculation, the mind was made susceptible to abstract generalization,' he claimed.[71] In his essay on Weber's concept of rationality, he writes that 'abstract reason becomes concrete in the calculable and calculated *domination* of nature and man . . . [and] reason envisaged by Weber thus is revealed as *technical* reason'.[72] Weber's idea of rationality was not value-free, Marcuse argues, but identifies 'technical reason with bourgeois capitalist reason'. Weber's failure to understand the way in which this idea of reason embodied a particular form of domination led him to erroneously argue that '"pure", formal, technical reason . . . erects the "shell of bondage"', rather than attributing cause to the historically specific 'reason of domination' that Marcuse associates with capitalism.[73] Marcuse thus connects science, technology, capitalism and domination – indeed, in today's world, these blend into one another. 'Today, domination perpetuates and extends itself not only through but *as* technology.'[74] The supposed neutrality of science is a dangerous myth, Marcuse emphasises; rather, 'the technological system is a system of domination which operates already in the concept and construction of techniques'.[75]

Enlightenment, Reason and Nature

In the eighteenth century, an emerging idea of reason already evident in this narrative of the scientific revolution is seen to develop into a more potent social and political force. Here, rationalisation prefigures modern technology and plays a key role in the development of 'technological thinking'. Enlightenment reason, these thinkers argue, encompasses a new faith in human progress and the progressiveness of human history, as well as positioning reason in a dominant relationship over nature. Both of these qualities are essential to the concept of technology that the critics develop, but it is particularly the transformed understanding of the relationship between humans and nature that shapes ideas of the technological.

In the *Dialectic of Enlightenment*, Adorno and Horkheimer claimed that Enlightenment reason is domination, of both nature and man. For Adorno and Horkheimer, writes Ferrone, the 'tight dialectical process that reversed the relationship between man and technology was all already present in the initial core of the Enlightenment's very way of thinking, which was bent on "establishing a unified, scientific order"'.[76] In his work on Hegel, Adorno writes of how 'rational science' brought about the congealing of philosophical consciousness: 'the unity of reification, that is, of a false – in Hegel's terms, abstract – objectivity external to the thing itself, and a naivete that confuses facts and figures, the plaster cast of the world, with its foundation'.[77] As this took place, and reason 'became autonomous and developed into an apparatus', Adorno claims, 'thinking also became the prey of reification and congealed into a high-handed method'.[78]

> Cybernetic machines are a crude example of this. They graphically demonstrate to people the nullity of formalized thinking abstracted from its

contents insofar as such machines perform better than thinking subjects much of what used to be the proud achievement of the method of subjective reason. Should thinking subjects passionately transform themselves into the instruments of such formalization, then they virtually cease being subjects. They approach the machine in the guise of its imperfect replica.[79]

Marcuse agrees with Adorno and Horkheimer, emphasising to an even greater degree the specifically technological character of modern politics and political domination. 'The totalitarian universe of technological rationality is the latest transmutation of the idea of Reason,' he states.[80] Today, technology has come to project its own world, argues Marcuse.[81] As such, 'technology' describes a mode of production, a 'totality of instruments', as well as a 'mode of organizing . . . social relationships . . . [and] an instrument for control and domination'.[82] Technique increasingly becomes '"pure" instrumentality', casting aside its original purpose of easing man's labours – its intended end is the development or expansion of technology itself, it is instrumental for technique, not man.[83] However, it is important to note that this is expressly *modern* technology, and Marcuse's historical analysis makes a case for a clear shift between the pre-technical age and the era of technology we now live in, with a form of repression that is essentially 'different from that which characterized the preceding, less developed stages of our society'.[84] Marcuse echoes Horkheimer and Adorno when he intimates that, although contemporary technological advance began in an ethos of individuality and rationality, these traditional standards 'are being dissolved by the present stage of the machine age'.[85] Specifically, 'individualistic rationality has been transformed into technological rationality. It is by no means confined to the subjects and objects of large scale enterprise, but characterizes the pervasive mode of thought and even the manifold forms of protest and rebellion.'[86]

Arendt also highlights some key distinctions between contemporary technology and the eighteenth century psyche. 'Modern advocates of automation', she points out, 'usually take a very determined stand against the mechanistic view of nature and against the practical utilitarianism of the eighteenth century, which were so eminently characteristic of the one-sided, single-minded work orientation of homo faber.'[87] At this time, she suggests, the tools of technology still served primarily to construct a world; they had not yet been turned to easing the labour of living; to automate the life process itself and thus unleash it upon and against the durable political and social world.

Marcuse characterises the technologisation of society as the '"denaturing" of reality', but points out that 'it is through man's own practices' that this has taken place.[88] Nature itself and 'human nature' are overcome and reshaped by technology in the modern world. The relationship between nature and technology is central to Marcuse's characterisation of technological development because, as Dana Villa writes, he is 'absolutely convinced that the domination of nature through technology is inextricably linked to the domination of man'.[89] This shapes the experience of man, and his experience of the world he lives in, totally and completely. 'Basic experience is no longer concrete experience, or social practice taken as a whole,' explains Marcuse, 'but administrative practice organized by technology. Such an evolution reflects the transformation of the natural world into a technical world. Technology, strictly speaking, has taken the place of ontology. The new mode of thought has cancelled the ontological tradition.'[90] That is, because our experience of the world is now mediated through technology, and because of the total nature of this, technology blocks access to the ontological reality of world and existence. Technology has replaced reason *and* nature in its never-ending quest for greater domination.

Adorno, Horkheimer and Marcuse argued that instrumental reason was complicit in a return to barbarism that catalysed in the twentieth century, irrevocably undermining faith in progress. Ellul and Mumford also locate the growth of modern technology, specifically technology in a systematised form, in the eighteenth century. Langdon Winner argues that, for Ellul, the Enlightenment and the French Revolution were 'masks for a much more fundamental and consequential movement: the technical revolution'.[91] But his analysis is more nuanced. The 'technical phenomenon' is 'specific to Western civilization since the eighteenth century', Ellul claims.[92] Enlightenment ideals of 'consciousness, criticalness, rationality' are pivotal to the emergence of this 'technological phenomenon', but, as a static phenomenon, technology had not yet reached its dynamic character as a system.[93] The turning point, for Ellul, seems to be the French Revolution and through it 'the emergence of a state that was truly conscious of itself and was autonomous in relation to anything that did not serve its interests', encompassing the 'economic technique' and 'rationalized systems' that characterise the modern technical state.[94] This state claimed, for the first time, to be the guarantor of liberty, he wrote, and the identification of the state's institutions with liberty would evolve into liberalism.[95]

For Mumford, the ideological structure of the modern, technological megamachine was realised in a widespread manner over the period from the late eighteenth century to the present day.[96] The 'reassemblage' and enhancement of the ancient machine began – reflecting a similar chronology to Ellul – with the French Revolution of 1789 and the birth of the 'National State, to which, on Rousseau's pseudo-democratic theory of the General Will, it bestowed absolute powers'.[97] Enlightenment 'ideals' served only to ease the megamachine's reassembly, he suggests; 'those who regarded faith in Progress as a moral imperative mistakenly assumed . . . that the forward motion was continuous, inevitable, and in most respects benign.'[98]

The Industrial Revolution and the Birth of Modernity

'For the most part,' Arendt proclaimed, 'we still live in a world defined by the industrial revolution,' while Jonas writes that 'modern technology, in the sense which makes it different from all previous technology, was touched off by the industrial revolution'.[99] Although the ideas which had generated the technologies of the Industrial Revolution had arisen in earlier times, with the birth of industry, the world as we recognise it began to take shape. The Industrial Revolution, while marking the origin of technological modernity in one sense, also represents the maturing of existing trends. The Industrial Revolution was a revolution in production, which could not be wholly understood in terms of the advancement of capitalism. Technology has its own, distinct path of development, these theorists all suggest, albeit there is a mutual influence between capitalist and technological development. These ideas connect the seemingly rather different perspectives on the Industrial Revolution in the writings of the critics of technology.

Heidegger's idea of technology as 'revealing' or 'enframing' is rooted in an idea and critique of industrial manufacturing.[100] But 'the revealing that holds sway throughout modern technology does not unfold into a bringing-forth in the sense of *poesis*', he writes; that is, it does not result in an authentic bringing-forth but rather, 'the revealing that rules in modern technology is a challenging, which puts to nature the unreasonable demand that it supply energy that can be extracted and stored as such'.[101] Thus, 'air is now set upon to yield nitrogen,' he argues, 'the earth to yield ore, ore to yield uranium, for example; uranium is set upon to yield atomic energy.'[102] The world becomes standing-reserve. Jonas agrees, writing that the 'first distinctive feature of modern technology . . . [is] the use of artificially generated and processed natural forces for the powering of work producing machines'.[103]

Although humans themselves in such a world are not swept *into* the 'standing-reserve', according to Heidegger, they are swept up with the misunderstanding or mis-ordering of the world they take part in by actively undertaking the work of technology, a work which is, he argues, not within human control.[104] As such, a 'merely instrumental' definition of technology is untenable, he argues.[105] The misunderstanding of technology runs deep in contemporary human self-understanding of being. As a result of technology's mode of enframing:

> *precisely nowhere does man today any longer encounter himself, i.e., his essence.* Man stands so decisively in attendance on the challenging-forth of Enframing that he does not apprehend Enframing as a claim, that he fails to see himself as one spoken to, and hence also fails in every other way to hear in what respect he ek-sists, from out of his essence, in the realm of an exhortation or address, and thus *can never* encounter only himself.[106]

Zimmerman summarises that, for Heidegger, it is:

> the contemporary *mode of understanding or disclosing things* which makes possible both industrial production processes and the modernity worldview. Both industrialism and modernity are symptoms of the contemporary disclosure of things as raw material to be used for expanding the scope of technological power for its own sake.[107]

The very abstract conceptualisation of technology as enframing takes on in industrialism and modernity more tangible forms: of economy and politics. It is, quite clearly, a critique of capitalism, in Heidegger's critique of world as 'standing-reserve', now the 'ground itself and . . . horizon' of our world. 'Only modern technology makes possible the production of all these economic standing-reserves,' he writes.[108]

Arendt offers her own distinct critique of technological capitalism in the industrial era. Capitalism, and the rise of the labouring activity (that is, those activities that are directed primarily to produce consumption goods), had resulted in a 'society of laborers': a society in which labour is deemed the essential and foremost activity. Yet the automation that has taken place through technological advances forms a kind of 'substitute for the real world', she argues.[109] The problem is twofold, in her opinion. First, the rise of labour itself as a politicised activity poses a threat to authentic political activity. Labour is necessary, cyclical action, and as such cannot be considered an activity of freedom, or the basis of free politics, nor can such an activity constitute a durable 'world', because of the transient nature of the things that labour produces. That is, modern politics, prioritising the management of the national economy or the organisation of labour, simply evades and obscures what is genuinely important about the political. But the rise of automation since the Industrial Revolution, in the society of labourers, removes the final sense of purpose from the citizens of the modern society; it reduces its human inhabitants to an irrelevance:

> Precisely because the animal laborans does not use tools and instruments in order to build a world but in order to ease the labors of its own life process, it has lived literally in a world of machines ever since the industrial revolution and the emancipation of labor replaced almost all hand tools with machines which in one way or another supplanted human labor power with the superior power of natural forces . . . Even the most primitive machine guides the body's labor and eventually replaces it altogether.[110]

Mumford agrees that the transformations of the industrial era should not be considered purely in terms of the technologies that emerged, arguing that underlying and pre-existing modes of technical organisation produce its technological artefacts. 'For the last century, even before Arnold Toynbee coined the term "the Industrial Revolution," the whole history of modern technology was misinterpreted by the Victorian over-valuation of the mechanical inventions of the eighteenth century and after,' he asserts.[111] An early modern system of 'one-sided political and military domination' came to produce 'its counterpart in a system of mechanization and automation that ignored the human premises upon which the older agricultural and handicraft technologies had been founded':[112]

> If craftsmanship had not been condemned to death by starvation wages [etc.] . . . our technology as a whole, even that of 'fine technics' would have been immensely richer – and more efficient . . . Whatever the advantages of a highly organized system of mechanical production, based on non-human sources of power – and as everyone recognizes there are many advantages – the system itself tends to grow more rigid, more inadaptable, more dehumanized in proportion to the completeness of its automation and its extrusion of the worker from the process of production.[113]

The new 'way of life' that the new industrial complex offers supposes that 'man has only one all-important mission in life', writes Mumford, 'to conquer nature . . . to remove all natural barriers and human norms and to substitute artificial, fabricated equivalents for natural processes'. Accordingly, he argues, there "is only one efficient speed, *faster;* only one attractive destination, *further away;* only one desirable size, *bigger;* only one rational quantitative goal, *more*'.[114] Capitalism conditioned the growth of technology (and vice versa), but Mumford also highlights the importance of military organisation.[115] 'Military regimentation proved the archetype for collective mechanization,' he argues, 'for the megamachine it created was the earliest complex machine of specialized, interdependent parts, human and mechanical.'[116]

Likewise, Ellul writes, in the eighteenth century technical progress would 'suddenly explode in every country and in every area of human endeavour'.[117] The Industrial Revolution was only part of this story, and Ellul criticises Mumford's (earlier) division of the industrial era by types of power (hydraulic, coal and electric) because it supposes technique to be equivalent to the machine. This is not the case, insists Ellul.[118] The technical revolution – the emergence of technological modernity –meant 'the emergence of a state that was truly conscious of itself and was autonomous in relation to anything that did not serve its interests'. This product of the French Revolution encompassed 'a precise military strategy . . . economic technique . . . rationalized systems . . . the great systematization of law . . . the unification of legal institutions . . . the translation into action of man's concern to master things by means of reason'.[119] It was not a purely technical shift, but a civilisational change in attitude.[120] Ellul identifies five reasons for this: the prior slow advancement of technological invention; population expansion; economic stability and growth; 'the plasticity of the social milieu', by which Ellul means the disappearance of traditional social groups and the emergence of individualism; and a 'clear technical intention' originating in industry's pursuit of efficiency and the 'one best way to do work'.[121] The rise of technique was the result of the coalescence of these five factors.

For the Frankfurt School, Aronowitz explains:

> mastery became domination when reason lost its distance from industrial technology. In essence, technical rationality meant the subordination of all natural problems to questions of social control. The machine became the chief mediation between human beings and nature by extending their productive powers to transform.[122]

Hence, Horkheimer writes, in industrial society thinking 'has to come down to earth'. Autonomy, intellectual freedom and innovation might have been valuable and valued at the dawn of the Enlightenment, but the success of the scientific project undermined this, as technology came to the fore. 'With the victory of technology . . . with men's control over nature . . . their autonomy, autonomy regresses, negates itself. What is under way in the bourgeois era will be completed in the automatized world. As the subject is being realized, it vanishes.'[123] This is the central claim of Horkheimer's *Eclipse of Reason:*

> [I aim] to inquire into the concept of rationality that underlies our con-temporary industrial culture, in order to discover whether this concept does not contain defects that vitiate it essentially . . . It seems that even as technical knowledge expands the horizon of man's thought and activity, his autonomy as an individual, his ability to resist the growing apparatus of mass manipulation, his power of imagination, his indepen-dent judgment, appear to be reduced. Advance in technical facilities for enlightenment is accompanied by a process of dehumanization. Thus progress threatens to nullify the very goal it is supposed to realize – the idea of man.[124]

Reason itself, although its contemporary importance originated in the ideals of the Enlightenment – of freedom and the realisation of man – has become wholly instrumentalised in modernity. 'Its role in the domination of men and nature has been made the sole criterion . . . The more ideas have become automatic, instrumentalized, the less does anyone see in them thoughts with a meaning of their own. They are considered things, machines.'[125] This comes about through industrialisation, whereby the 'factory is the prototype of human existence'.[126]

Technology, for Horkheimer, is not the only contributing factor to the contemporary domination of man, but it is increasingly the form that all dominations become centred in and realised through: the economic domination of modern societies, the rationalisation (thus de-rationalisation) of individuals, and the mastery of nature. Its evolution to full automation only entrenches this fact. Yet, 'the fault is not in machines', Horkheimer argues. The machine 'points to a legitimate condition of mankind'.[127] The problem is the way that machines have been co-opted and drawn into a system of domination, whereby technology can thus provide 'a new dimension to productive and destructive force'.[128] Technology might be the salvation or ruin of society, but, Horkheimer indicates, its utilisation, construction and systematisation in modernity draws it to the latter.

THE ATOM BOMB AND WORLD WAR: TECHNOLOGY AS TOTALITARIANISM

The birth of a new era of is marked by the arrival of the atom bomb. Yet, in this narrative, the bomb is not an unexpected development, but rather a continuation of already powerful trends. The bomb disrupted an already precarious balance between human processes of production and destruction – forces which had, through technology, already been changing and developing at a rapid pace in modernity. Unlike earlier technologies, writes Arendt, in a 'technology driven by the processes of nuclear energy . . . processes that do not occur naturally on earth are brought to earth to produce a world or destroy it'.[129] The atom bomb reflects what for Arendt constitutes the essential characteristic of contemporary technologies: that rather than being driven by a more straightforward drive for an improvement in methods of production and efficiency, they release 'natural forces' into the human world. 'The chain reaction of the atom bomb, therefore, can easily become the symbol for a conspiracy between man and the elementary forces of nature, which, when touched off by human know-how, may one day take their revenge and destroy all life on the surface of the earth, and perhaps even the earth itself.'[130] The bomb is an existential threat because it brings the forces of nature within the grasp of humans for the first time, although not within our control.

The atom bomb, therefore, while situated within the developmental trajectory of technology, is at the same time much more than the regular development of technologies of warfare; it is a leap into a new kind of technology and a new kind of world. The 'development, possession and threatened use of atomic weapons . . . is a primary fact of political life'.[131] It threatens not just life, but the very life-world of societies and their values: in the atomic age, 'both the Hebrew-Christian limitation on violence and the ancient appeal to courage have for all practical purposes become meaningless, and, with them the whole political and moral vocabulary in which we are accustomed to discuss these matters'.[132] The atom bomb's potential completes a collapse of traditional values and ways of understanding the world, with little else to replace them. The merging of the human world, via technology, with nature, undermines humanistic principles, she believes. As Canovan points out, 'in [Arendt's] interpretation, totalitarianism involves treating the human world as if it were a part of nature.'[133] The modern change in the relationship between nature and the human world is, for all of our thinkers, anything but benign. Technology is an attempt to enlist the forces of nature for the purposes of domination, and to dominate nature itself. This is central to the character of modern technology. But Arendt highlights the impossibility of this task: that through our attempts to harness nature's forces, we extend and increase its disruptive and destructive power over the human world.

Arendt argues that the introduction of nature into the human world, or the attempt to harness natural forces, has powerful political implications. The political world is a world of human construction, she writes: an artifice. This world offers human beings some durability and stability in a world of natural flux, and it offers individuals the space into which to act, to be freed from the sheer necessity of existence, to act in more meaningful ways within the space of a wider community. The channelling of nature into the political world threatens this stability as the excessive forces of nature run the risk of spinning out of control. Furthermore, the 'channeling of natural forces into the human world has shattered the very purposefulness of the world', Arendt writes, 'the fact that objects are the ends for which tools and implements are designed.'[134] Human tools, augmenting and enabling the processes of work which construct our world, are used towards some end, based on some model. Natural processes do not operate on this basis, but are never-ending. Thus technologies that attempt to harness natural forces only accelerate and exacerbate the risks of human action, and undermine the process of work which alone can create the durable world that in prior times helped to restrain and direct action. The question of technology, therefore, 'is not so much whether we are the masters or the slaves of our machines, but whether machines still serve the world and its things, or if, on the contrary, they and the automatic motion of their processes have begun to rule and even destroy world and things'.[135] The 'pseudo world' of machines 'cannot fulfil the most important task of the human artifice, which is to offer mortals a dwelling place more permanent and more stable than themselves', she writes.[136] Instead, 'the rhythm of machines would magnify and intensify the natural rhythm of life enormously, but it would not change, only make more deadly, life's chief character with respect to the world, which is to wear down durability'.[137] As Simbirski writes, for Arendt:

> the 'science' of the atomic age was essentially technological, having particular bearing on politics. Technology did not simply follow from scientific

> progress . . . technological transformation had become the essential, defining characteristic of the era . . . [and] assuming the sustained acceleration of technological change through automation and the reduction of science to a part of that process.[138]

Anders argues that modern man has become 'homo creator' because we are capable of generating 'products from nature that do not belong to the class of "cultural products" . . . but rather to the class of nature . . . there are now processes and pieces of nature that did not exist before they were created by us'.[139] The advent of nuclear power has stopped the passage of history, he argues, because the risk embodied in nuclear weapons is so great that it can never be eradicated. We live, therefore, 'in an age in which we are continuously producing our own destruction (the only thing we do not know is its precise moment)'.[140]

Mumford claimed that both ancient and modern megamachines were essentially totalitarian, utilising a rather different and more capacious idea of totalitarianism to most of the other theorists here. Arendt argued totalitarianism was a novel and distinctively modern regime; Anders wrote that the term 'totalitarian' refers to a political situation that has 'fundamentally changed'; Marcuse connects totalitarianism to 'late industrial civilization' and the influence of the interests of domination on productivity.[141] Mumford, however, argues that totalitarian and democratic tendencies (centralised versus decentralised; 'the mass' versus small communities) have always been present in the technology of societies ancient and modern. Yet he believes that modern technology distinguishes the modern 'megatechnic complex' from the older structures composed of human material. The 'modern machine has progressively reduced the number of human agents and multiplied the more reliable mechanical and electronic components', he explains. This 'not merely reduc[es] labor force needed for a colossal operation but facilitate[es] instantaneous remote control . . . The modern machine escapes temporal and spatial limitations.'[142] The replacement of organisms by machines has 'elevated the mechanical creature above his creator . . . [an] error [which] has brought catastrophic potentialities in our day . . . to give to agents of extermination . . . – nuclear weapons, rockets, lethal poisons and bacteria – the authority to exterminate the human race'.[143] Like Arendt, Mumford sees in the era of total war not just the capacity for total annihilation, but the tendency for technology to relegate human beings to a lesser role, even an irrelevant one.

Jonas writes on Hiroshima's and Nagasaki's destruction, of how this 'shock, perpetuated by the immediately following nuclear arms race, was the first trigger for a new and anxiety-ridden rethinking of the role of technology in the Western world'.[144] Technology threatens the human world and natural environment in which we live, not just through the destruction that ensues deliberately from the utilisation of nuclear weapons, or even the associated (but less deliberate) nuclear fallout, for example, but more insidiously, through the unintended consequences of technology such as pollution and the overuse of natural resources. Nature shows a 'critical vulnerability . . . to man's technological intervention,' writes Jonas in the early 1970s, 'unsuspected before it began to show itself in damage already done'.[145]

Technology has transformed the nature of human beings because, with the development of technological power, human action itself has been transformed, Jonas argues.[146] 'Man himself has been added to the objects of technology. *Homo faber* is turning upon himself and gets ready to make over the maker of all the rest.'[147] Thus we find again the equating of the control of nature with the domination of man, and the remarkable decline in the autonomous power of *men* at the same time as *man* becomes ever more powerful.[148] Because modern technology bears the constant threat of global or total catastrophe, it also collapses ethical understanding. It has 'eroded the foundations from which norms could be derived; it has destroyed the very idea of norm as such,' says Jonas.[149] The great changes that are taking place, and the possibility of catastrophe, necessitate a scepticism towards technology, he argues. Modern technology 'renders obsolete the tacit standpoint of all earlier ethics that, given the impossibility of long-term calculation, one should consider what is close at hand only and let the distant future take care of itself'.[150] This ethics does not serve the current state of humanity. Instead, we must recognise that 'care for the future of mankind is the overruling duty of collective human action in the age of a technical civilization that has become "almighty," if not in its productive then at least in its destructive potential'.[151] We should consider our ultimate ethical and political task to mitigate and limit the technological risks to our future. As such, the problems of political rule have been magnified exponentially by the 'magnitude and complexity (also sophistication) of the forces to contend with', as has the problem of avoiding that power being 'beholden to the interests which the technological colossus generates on its path'.[152]

For Arendt, Mumford and Jonas, technology transforms and diminishes human action. For Heidegger, it is the representation of a trend transforming (and undermining) thought. But they agree that nuclear technology represents the culmination of a much longer development. 'It is not as a particular deadly machine that the much discussed atom bomb is deadly,' Heidegger writes:

> What has long threatened man with death, indeed with the death of his essence, is the absoluteness of his sheer willing in the sense of his deliberate self-assertion in everything . . . What threatens man in his essence is the opinion that technological production would bring the world into order, when it is exactly this ordering that flattens out each *ordo*, that is, each rank, into the uniformity of production and so destroys in advance

> the realm that is the potential source from which rank and appreciation originate out of being.[153]

What was done to Hiroshima and Nagasaki also highlighted another feature of technological modernity: the willingness of liberal-democratic, 'free' political regimes to utilise the most devastating and total destructive technologies in order to attain their ends. The total mobilisation of much of the developed world's resources towards destruction in the First and Second World Wars revealed the willingness of liberal states to use any technological means; the use of atomic weapons reinforced this absolutely. From different political positions, Mumford, Heidegger, Ellul, Arendt, Jonas, Anders and the Frankfurt School thinkers all make the claim that technology's impact on modernity is equally central to West and East, whether capitalist or communist, democratic or fascist. Technology may be realised in different ways, but the essence of it, and the totalitarian direction of technology is the same worldwide (a theme that the following chapter will pick up in more detail). Technology provides a structural identification for all developed states in the post–Second World War world.

Mumford, for instance, highlights, as a key step in the reassembly of the megamachine, the political and social changes that resulted from the First World War, including: 'the enlistment of scholars and scientists as an arm of the state, and the placating of the working classes by universal suffrage, social welfare legislation, national elementary education, job insurance and old-age pensions'.[154] The next step, however, was taken by the Soviets and Nazis, whose support, he notes, might have made one think the megamachine would be completely discredited. Far from it, the West took over the reconstruction of the megamachine themselves, in their pursuit of weapons that might counter their enemies.[155] 'In the 1930s and beyond,' Zimmerman writes, 'Heidegger interpreted Americanism and Bolshevism as different versions of the nihilism of modern technology – nihilism which could be overcome only by a complete transformation of German *Dasein*.'[156] Despite Heidegger's support for the Nazis in the 1930s (and arguably beyond), he also placed Germany within this framework, when he 'chose to view the Holocaust as a *typical* episode in the technological era afflicting the West'.[157] Like Heidegger, Horkheimer merges the opposing forces of the Second World War: the Anglo-Saxon world has moved to an alliance with 'the like' of Nazism across the world.[158] Meanwhile the 'communist' countries are no closer to enabling 'the free development of human powers'.[159] The state of society, individuality and politics is the same the world over, and that is thanks to technology. 'The individuals in the West come to resemble those of the East. Technology has made superfluous the freedom which its development required.'[160] Adorno claims that 'progress and barbarism are today . . . matted together in mass culture'. Liberalism 'hatched the culture industry', Adorno argues, but was also 'the ancestor of the fascism which destroyed both it and its later potential customers'.[161] Dictatorship did not 'swoop down' upon our societies, he insists, but were 'engendered by the social dynamic following the First World War and they cast their shadows behind. This is immediately obvious in the products of mass culture manipulated by a highly centralized economic power.'[162]

Ellul identifies Lenin as the progenitor of the first truly political technique.[163] Yet, despite his critique of communism (and equally fascism), he identifies the same trends occurring within the West.[164] In today's world, he writes, 'people are no freer on one side than the other [of East and West]; they are just placed into the service of production by different means.'[165] Technique, beginning in the Enlightenment and Industrial Revolution of the West, has become a global phenomenon – *the* overwhelming global phenomenon. 'Without exception in the course of history, *technique belonged to a civilization . . .* today *technique has taken over the whole of civilization.*'[166] This is the West's 'greatest crime', he argues, that 'our technical society is not a true expression of the West but its betrayal'.[167]

Despite the importance of the atom bomb as a moment when technology's true nature and the extent of its power is visible, technology's character, as we have seen, means that it operates as much in peacetime as war. Technology exerts dominance through its various methods of destruction but also through psycho-cultural means.

The development of reason, in a technological or mathematicised form, and through industrial technology in the modern era has shaped the individual, society and politics in terrible ways, Horkheimer argues. The human psyche is changing, to the point where the 'awareness of the self as an autonomous individual with his own soul is giving way to the corporate mentality'.[168] The contemporary individual has 'only enough spontaneity to launch himself onto the path prescribed for him', in other words, none at all.[169] 'Interiority has withered away', to be replaced with 'technological expertise, presence of mind, pleasure in the mastery of machinery, the need to be part of and to agree with the majority or some group . . . whose regulations replace individual judgment'.[170] Men today act more and more like machines, he writes.[171] The disappearance of the individual results in the disappearance of society, each the essential antagonist to one another. A 'better social totality' can only emerge 'when our conscience refuses to rest easy with the disappearing freedom of the individual'.[172] This, of course, has radically shaped modern politics: today democracy is an 'illusion'; liberalism and authoritarianism cannot be distinguished (and even support one another), or, put another way, in its Enlightenment form, political liberalism has been extinguished.[173]

Technology, claims Adorno, disrupts normal social interaction through the replacement of true culture with mass culture, and enables the total domination of masses through technologies such as the television. The 'deceit' of the culture industry, he explains, and its great success:

> consists precisely in confirming and consolidating by dint of repetition mere existence as such, what human beings have been made into by the way of the world . . . The vocabulary of the image-writing is composed of stereotypes. They are defended with technological imperatives, such as the need to produce in a minimal time a terrific quantity of material.[174]

The false immediacy brought about by mass culture, he writes, 'an identity created by the expansion of technology, amounts to the affirmation of the relations of production . . . [and this] system has now become independent, even of those who are in control'.[175]

TECHNOLOGY AND THE FUTURE

The narratives of the development of technology outlined here are clearly not identical. Different thinkers place more or less emphasis on different moments in their histories; they interpret the influence of historical events in different ways which speak to their own wider theoretical frameworks or claims. Their chronologies of the origins of modern technologisation differ, with Heidegger, Mumford, Ellul and Adorno and Horkheimer identifying the beginnings in different parts of the ancient world, while Arendt and Jonas argue the ancient Greeks held technology at a distance precisely because they understood the threats it posed. They are at odds on the influence of Christianity: to Heidegger, the rise of Christianity had little impact on the rise of technology; to Ellul it restrained it; for Mumford, Christianity both restrained technology *and* added new elements to technology that would later become influential. These discrepancies are important because they suggest a different trajectory of 'modernization'; that is, some of the critics of technology argue that technology's origins can be traced back to the pre-modern era. From the seventeenth century their narratives draw closer together. The Scientific Revolution, the Enlightenment, and the Industrial Revolution represent, respectively, the emergence of 'technology' in science, its extension into philosophy in the form of instrumental reason, and its materialization into the mass technologies of contemporary society. With the atom bomb, the critics are clear that technology has reached a new stage, in which humanity has come to challenge nature itself on a global scale. There are therefore two significant relationships here which are more important than any other: the relationship between science and technology, and the relationship between nature and technology. For both, these critics of technology make extremely similar claims.

Modern technology, they argue, is preordained by modern science, or more accurately, they argue that science was technological from the start.

They highlight the instrumentality of modern science, both understood in terms of an instrumental rationality, but also a literal instrumentality: modern science relies upon a physical instrumentarium. This emphasis on the physical instrumentalism of technology is important in distinguishing 'technology' from rationality *alone* – technology is not only science, or rationalisation. These things change the way that humans and human knowledge relate to the world; we become technological through science because science is inherently technological. The claim is highly deterministic. Once modern science enters the world, it seems almost impossible that humanity could travel in any other direction.

The way technology changes humanity's relationship with nature is another central theme in the development of technology, perhaps even the most important, and one which will arise again and again in the chapters that follow. Technology is described as an attempt to dominate nature, an attempt to chain nature (and thus release natural forces), an attempt to change nature and, on this basis, also an attempt to dominate, enslave or transform humanity. It is not the case for these thinkers that what is 'natural' is necessarily 'good'; the unnatural or artificial can, in the right context, be perfectly acceptable, and even a necessary aspect of political and social existence. What is significant in the relationship between technology and nature as it unfolds in modernity is the attempt to transform nature from something that is simply an unchangeable environment, in the sense that it is beyond our control, or that it represents a power beyond us, into something that is within human control and may be comprehensively understandable and manipulated. Nature, in such a reframing, is at the mercy or whim of humans; we become as gods and nature is there simply to be used.

The problem with this, as it appears in the work of these theorists, is twofold. In one sense, although these thinkers do not, on the whole, idealise nature (though aspects of idealisation are apparent at times), there is certainly a sense that nature is ungraspable and uncontrollable. It exists, and attempts to manipulate it will invariably fail or go awry. On the other hand, these thinkers recognise all too clearly the incredible power of modern technology and the way that technology can and does manipulate natural processes. We (and who 'we' are is, of course, important) are *sometimes* in control of this extraordinary power, but never fully in control. The effective limits on our control of technology, or even our understanding of how we are manipulating nature, may have terrible consequences. In other words, human influence over nature via technology represents hubris. It is either doomed to go wrong or, even when successful, its impact can be catastrophically dehumanising.

What, then, does the future of technology hold for us? There is little chance of technology overcoming itself. Even Marxists, in the twentieth century, became deeply sceptical about Marx's optimism about technology in a communist future. It was among Marx's greatest mistakes, writes Horkheimer, to suggest that 'consciousness would become free once men mastered material conditions; to imagine that peace among the classes would "also be peace among men and with nature"'.[176] Ellul and Jonas both specifically attack ideas of technological utopia. 'Utopia lies in the technological society, within the horizon of technology. And nowhere else,' writes Ellul.[177]

> What the modern utopist unwittingly creates is a radically technicized world! Only the visible, striking drawbacks of technique have been eliminated; in reality the utopia represents, in the guise of a dream, the unqualified triumph of technical rationalism. The supposedly revolutionary imagination produces, in fact, an idea that is as antirevolutionary as anything could possibly be.[178]

Projecting forward, Arendt suggests that, 'If present technology consists of channeling natural forces into the world of the human artifice, future technology may yet consist of channeling the universal forces of the cosmos around us into the nature of the earth.'[179] Heidegger, in a similar vein, writes that man seeks to 'arrogate to himself the hidden working of nature in the form of energy, and to subordinate the course of history to the plans and orderings of a world government'.[180] Citing the pursuit of genetic and medical technologies, Arendt argues that 'future man . . . seems to be possessed by a rebellion against human existence as it has been given, a free gift from nowhere (secularly speaking), which he wishes to exchange, as it were, for something he has made himself'.[181]

Nevertheless, although these theorists are in the main (on this topic) convinced on the very likely continuation of the technological trends they identify, they are not *wholly* deterministic. What is required for a meaningful change in the impact of technology on society is a revolutionary change in both technology and society, interconnected as they currently are. Even so, there is still hope. Of the thinkers whose work is explored here, Marcuse, though explicitly making the connection between technology and domination, and centring its overwhelmingly destructive qualities and capacity in his work, is perhaps the most optimistic about technology's future. Certainly, humanity cannot go back to a pre-technological stage, but he suggests there might be possibilities for using the 'achievements of technological civilization for freeing man and nature from the destructive abuse of science and technology in the service of exploitation' – for using technology against itself.[182] This would necessitate the overcoming of the entire social structure that supported and required contemporary technology, but for Marcuse, more hopeful about the possibility of revolution than either Adorno or Horkheimer, this was not necessarily a bar.

Mumford's optimism comes from a different direction. Only one thing can oppose the megamachine and its totalitarian mode of existence, he argues: the organic. Organisms, unlike mechanisms, are not closed but 'open system[s], subject to chance mutations and to many external forces and circumstances over which it has no control'.[183] An 'organic system' might oppose the emphasis of contemporary technological society on sheer abundance, he suggests, by directing itself to 'qualitative richness, amplitude, spaciousness, free from quantitative pressure and crowding, since self-regulation, self-correction, and self-propulsion are as much an integral property of organisms as nutrition, reproduction, growth and repair'.[184]

Jonas's *Imperative of Responsibility* suggests that only a revolution in ethical understanding can tackle technology's challenges:

> Care for the future of mankind is the overruling duty of collective human action in the age of a technical civilization that has become 'almighty,' if not in its productive then at least in its destructive potential . . . There is no need . . . to debate the relative claims of nature and man when it comes to the survival of either, for in this ultimate issue their causes converge from the human angle itself . . . the 'No to Not-Being' – and first to that of man, is at this moment and for some time to come the primal mode in which an emergency ethics of the endangered future must translate into collective action the 'Yes to Being' demanded of man by the totality of things.[185]

Thus, technology does not necessarily determine our future. Freedom is still possible. But the state of technology in modernity means that only radical, thorough-going change to the society that technology is now embedded in, has the power to transform technology's ongoing effect.

NOTES

1. Jeffrey Herf, '"Dialectic of Enlightenment" Reconsidered', *New German Critique* 117 (2012), 84.
2. David Edgerton, *The Shock of the Old: Technology and Global History since 1900* (London: Profile, 2008).
3. Martin Heidegger, *Parmenides* [1942] (Bloomington and Indianapolis: Indiana University Press, 1992), 86.
4. Martin Heidegger, 'The Question Concerning Technology', in *The Question Concerning Technology and Other Essays*, trans. William Lovitt (New York: Harper & Row, 1977), 13.
5. Martin Heidegger, *Four Seminars* (Bloomington and Indianapolis: Indiana University Press, 2003), 37.
6. Michael E. Zimmerman, *Heidegger's Confrontation with Modernity: Technology, Politics, Art* (Bloomington and Indianapolis: Indiana University Press, 1990), xv.
7. Gerard Delanty and Neal Harris, 'Critical Theory and the Question of Technology: The Frankfurt School Revisited', *Thesis Eleven* 166:1 (2021), 90.

8. Hannah Arendt, *The Human Condition* (Chicago: University of Chicago Press, 1998), 157.
9. Hans Jonas, *The Imperative of Responsibility* (Chicago: University of Chicago Press, 1984), 9.
10. Hans Jonas, 'Seventeenth Century and After: The Meaning of Scientific and Technological Revolution', in *Philosophical Essays* (New York and Dresden: Atropos Press, 1980), 80.
11. Jacques Ellul, *The Technological Society*, trans. John Wilkinson (New York: Random House, 1964), 27.
12. Ibid.
13. Ibid., 29.
14. Ibid.
15. Ibid., 30.
16. Ibid.
17. Lewis Mumford, *The Myth of the Machine: Technics and Human Development* (New York: Harcourt Brace Jovanovich, 1967), 11.
18. Ibid., 211.
19. Ibid., 190.
20. Ibid., 183.
21. Ellul, *The Technological Society*, 34.
22. Ibid., 39.
23. Zimmerman, *Heidegger's Confrontation with Modernity*, 171.
24. Arendt, *The Human Condition*, 313–14.
25. Lewis Mumford, *The Myth of the Machine: The Pentagon of Power* (New York: Harcourt Brace Jovanovich, 1970), 141.
26. Mumford, *The Myth of the Machine: Technics and Human Development*, 263.
27. Ibid., 263.
28. Lewis Mumford, *Technics and Civilization* (San Diego, New York and London: Harcourt Brace Jovanovich, 1963), 53.
29. Paul Forman, 'The Primacy of Science in Modernity, of Technology in Postmodernity, and of Ideology in the History of Technology', *History and Technology* 23:1/2 (2007), 1.
30. Andrew Feenberg, *Technology, Modernity and Democracy* (London and New York: Rowman & Littlefield, 2018), 67.
31. Edmund Husserl, *The Crisis of European Sciences and Transcendental Phenomenology* (Evanston: Northwestern University Press, 1970), 48.
32. Ernst Cassirer, 'Form and Technology' [1930], in *Ernst Cassirer on Form and Technology: Contemporary Readings*, ed. Aud Sissel Hoel and Ingvild Folkvord (Basingstoke: Palgrave Macmillan, 2012), 43.
33. Theodor W. Adorno and Max Horkheimer, *Dialectic of Enlightenment* [1944/47] (London: Verso, 1997), 19.
34. Ibid., 4.
35. Ibid.
36. Ibid., 11.
37. Ibid., 39.
38. Max Horkheimer, *Eclipse of Reason* (London: Continuum, 1947), 16.

39. Ibid., 72.
40. Adorno and Horkheimer, *Dialectic of Enlightenment*, 30; Horkheimer, *Eclipse of Reason*, 64.
41. Adorno and Horkheimer, *Dialectic of Enlightenment*, 42.
42. Arendt, *The Human Condition*, 11.
43. Ibid., 264.
44. Ibid., 231.
45. Ibid., 257–8.
46. Martin Heidegger, *What is a Thing?* [1935], trans. W. B. Barton Jr. and Vera Deutsch (Chicago: Henry Regnery Company, 1967), 65.
47. Ibid., 51.
48. Ibid., 93.
49. Ibid., 101.
50. Heidegger, 'The Question Concerning Technology', 22.
51. Ibid., 72.
52. Jonas, 'Seventeenth Century and After', 48.
53. Ibid., 72.
54. Ibid., 48.
55. Ibid., 49.
56. Ibid., 48.
57. Ibid.
58. Ibid., 48–9.
59. Hans Jonas, *The Phenomenon of Life* (New York: Harper & Row, 1966), 198.
60. C. Fred Alford, *Science and the Revenge of Nature: Marcuse and Habermas* (Gainesville: University Presses of Florida, 1985), 3.
61. Jacques Ellul, *The Presence of the Kingdom*, trans. Olive Wyon (Philadelphia: Westminster Press, 1951), 41.
62. Paul Forman, 'How Lewis Mumford Saw Science, and Art, and Himself', *Historical Studies in the Physical and Biological Sciences* 37:2 (2007), 273.
63. Mumford, *The Myth of the Machine: The Pentagon of Power*, 39.
64. Ibid., 57; 58.
65. Ibid., 57.
66. Ellul, *The Presence of the Kingdom*, 41.
67. Ibid.
68. Ellul, *The Technological Society*, 7.
69. Ibid., 129.
70. Ibid., 7.
71. Herbert Marcuse, *One-Dimensional Man* (Boston, MA: Beacon, 1964), 138.
72. Herbert Marcuse, 'Industrialization and Capitalism in the Work of Max Weber' [1964], in *Negations: Essays in Critical Theory* (London: MayFlyBooks, 1968), 154.
73. Ibid., 167–8.
74. Marcuse, *One-Dimensional Man*, 164.
75. Ibid., xvi.
76. Vincenzo Ferrone, *The Enlightenment: History of an Idea* (Princeton: Princeton University Press, 2015), 31.

77. Theodor Adorno, *Hegel: Three Studies* (Cambridge, MA: MIT Press, 1993), 73–4.
78. Theodor Adorno, 'Notes on Philosophical Thinking', in *Critical Models: Interventions and Catchwords*, trans. H. W. Pickford (New York: Columbia University Press, 1998), 127.
79. Ibid., 127–8.
80. Marcuse, *One-Dimensional Man*, 123.
81. Ibid., 154.
82. Herbert Marcuse, 'Some Social Implications of Modern Technology', *Studies in Philosophy and Social Science* 9 (1941), 414.
83. Herbert Marcuse, 'World without a Logos', *Bulletin of the Atomic Scientists* 20 (January 1964), 25.
84. Marcuse, *One-Dimensional Man*, x.
85. Marcuse, 'Some Social Implications of Modern Technology', 415.
86. Ibid., 417.
87. Arendt, *The Human Condition*, 151.
88. Marcuse, 'World without a Logos', 25.
89. Dana Villa, *Public Freedom* (Princeton: Princeton University Press, 2008), 162.
90. Herbert Marcuse, 'World without a *Logos*', *Bulletin of the Atomic Scientists* 20 (January 1964), 25.
91. Langdon Winner, *Autonomous Technology: Technics-out-of-Control as a Theme in Political Thought* (Cambridge, MA: The MIT Press, 1977), 124.
92. Ellul, *The Technological Society*, 23.
93. Ellul, *The Technological Society*, 23; Jacques Ellul, *The Technological System*, trans. Joachim Neugroschel (New York: Continuum, 1980), 79.
94. Ellul, *The Technological Society*, 43.
95. Jacques Ellul, *Autopsy of Revolution*, trans. Patricia Wolf (New York: Alfred A. Knopf, 1971), 83.
96. Mumford, *The Myth of the Machine: The Pentagon of Power*, 100.
97. Ibid., 245.
98. Lewis Mumford, *My Works and Days: A Personal Chronicle* (New York and London: Harcourt Brace Jovanovich, 1979), 12.
99. Hannah Arendt, 'Introduction *into* Politics', in *The Promise of Politics* (New York: Schocken Books, 1993), 154-5; Jonas, 'Seventeenth Century and After', 76.
100. Heidegger, 'The Question Concerning Technology', 12.
101. Ibid., 14.
102. Ibid., 15.
103. Jonas, 'Seventeenth Century and After', 76.
104. Heidegger, 'The Question Concerning Technology', 18.
105. Ibid., 21.
106. Ibid., 27.
107. Zimmerman, *Heidegger's Confrontation with Modernity*, xiii.
108. Heidegger, *Four Seminars*, 62.
109. Arendt, *The Human Condition*, 152.
110. Ibid., 146–7.
111. Mumford, *The Myth of the Machine: The Pentagon of Power*, 130.
112. Ibid., 142.

113. Ibid., 143.
114. Ibid., 172–3.
115. Mumford, *Technics and Civilization*, 26.
116. Mumford, *The Myth of the Machine: The Pentagon of Power*, 150.
117. Ellul, *The Technological Society*, 42.
118. Ibid., 42. However, Ellul does cite Mumford approvingly elsewhere in the same text, for example that Mumford wrote of the machine that it 'tends, by reason of its progressive character, to the most acute forms of human exploitation'. *The Technological Society*, 5.
119. Ibid., 43.
120. Ibid., 44.
121. Ibid., 47; 51; 52; 52–3.
122. Stanley Aronowitz, *Science as Power: Discourse and Ideology in Modern Society* (Minneapolis: University of Minnesota Press, 1998), 128.
123. Max Horkheimer, *Dawn and Decline* (New York: The Seabury Press, 1978), 229.
124. Horkheimer, *Eclipse of Reason*, 4
125. Ibid., 14–15.
126. Ibid., 35.
127 Max Horkheimer, 'The Concept of Man' [1957], in *Critique of Instrumental Reason* (London: Verso, 2012), 28.
128. Ibid.
129. Arendt, 'Introduction *into* Politics', 157.
130. Hannah Arendt, 'Europe and the Atom Bomb' [1953], in *Essays in Understanding 1930–1954: Formation, Exile, and Totalitarianism* (New York: Schocken Books, 2005), 419.
131. Ibid., 418.
132. Ibid., 422.
133. Margaret Canovan, *Hannah Arendt: A Reinterpretation of Her Political Thought* (Cambridge: Cambridge University Press, 1992), 81.
134. Arendt, *The Human Condition*, 150.
135. Ibid., 151.
136. Ibid., 152.
137. Ibid., 132.
138. Brian Simbirski, 'Cybernetic Muse: Hannah Arendt on Automation, 1951–1958', *Journal of the History of Ideas* 77:4 (2017), 606.
139. Günther Anders, *Die Antiquiertheit des Menschen*, vol 2: *Über die Zerstörung des Lebens im Zeitalter der dritten industriellen Revolution* [1980] (Munich: Verlag C. H. Beck, 2018), 22.
140. Ibid., 20.
141. Günther Anders, 'Theses for the Atomic Age', *The Massachusetts Review* 3:3 (1962), 494–5; Marcuse, *Eros and Civilization*, 93.
142. Mumford, *The Myth of the Machine: The Pentagon of Power*, 258.
143. Ibid., 97.
144. Hans Jonas, 'Prologue', in *Mortality and Morality: A Search for Good after Auschwitz*, ed. Lawrence Vogel (Evanston: Northwestern University Press, 1996), 49–50.

145. Hans Jonas, 'Technology and Responsibility: Reflections on the New Tasks of Ethics' [1972/3], in *Philosophical Essays* (New York and Dresden: Atropos Press, 1980), 9.

146. Ibid., 3.

147. Ibid., 14.

148. Hans Jonas, 'Toward a Philosophy of Technology', *The Hastings Centre Report* 9:1 (1979), 201.

149. Jonas, 'Technology and Responsibility', 19.

150. Jonas, *The Imperative of Responsibility*, 34.

151. Ibid., 136.

152. Jonas, 'Toward a Philosophy of Technology', 202.

153. Martin Heidegger, 'Why Poets?', in *Off the Beaten Track*, trans. Julian Young and Kenneth Haynes (Cambridge: Cambridge University Press, 2002), 221.

154. Mumford, *The Myth of the Machine: The Pentagon of Power*, 245.

155. Ibid., 250.

156. Zimmerman, *Heidegger's Confrontation with Modernity*, 42.

157. Ibid., 43.

158. Horkheimer, *Dawn and Decline*, 187.

159. Max Horkheimer, 'Foreword' [1967], in *Critique of Instrumental Reason* (London: Verso, 2012), ix.

160. Horkheimer, *Dawn and Decline*, 174.

161. Theodor Adorno, *The Jargon of Authenticity* (London: Routledge and Kegan Paul, 1973), 49.

162. Theodor Adorno, 'Those Twenties', in *Critical Models: Interventions and Catchwords*, trans. H. W. Pickford (New York: Columbia University Press, 1998), 41.

163. Ellul, *The Technological Society*, 232.

164. Jacques Ellul, *The Betrayal of the West*, trans. Matthew O'Connell (New York: Seabury, 1978).

165. Ellul, *The Presence of the Kingdom*, 21

166. Ellul, *The Technological Society*, 128.

167. Ellul, *The Betrayal of the West*, 17; 63.

168. Max Horkheimer, 'Threats to Freedom', in *Critique of Instrumental Reason* (London: Verso, 2012), 157.

169. Horkheimer, 'The Concept of Man', 4.

170. Ibid., 12.

171. Ibid., 26.

172. Horkheimer, 'Threats to Freedom', 139.

173. Adorno and Horkheimer, *Dialectic of Enlightenment*, 21; Horkheimer, *Eclipse of Reason*, 49; 105.

174. Theodor Adorno, 'Television as Ideology', in *Critical Models: Interventions and Catchwords*, trans. H. W. Pickford (New York: Columbia University Press, 1998), 55.

175. Theodor Adorno, 'Late Capitalism or Industrial Society?', in *Can One Live after Auschwitz?: A Philosophical Reader* (Stanford: Stanford University Press, 2003), 124–5.

176. Horkheimer, *Dawn and Decline*, 156.

177. Ellul, *The Betrayal of the West*, 20.
178. Ibid., 151.
179. Arendt, *The Human Condition*, 150.
180. Martin Heidegger, 'Anaximander's Saying', in *Off the Beaten Track*, trans. Julian Young and Kenneth Haynes (Cambridge: Cambridge University Press, 2002), 280–1.
181. Arendt, *The Human Condition*, 2–3.
182. Herbert Marcuse, *Counterrevolution and Revolt* (Boston, MA: Beacon Press, 1972), 60.
183. Mumford, *The Myth of the Machine: The Pentagon of Power*, 97.
184. Ibid., 395–6.
185. Jonas, *The Imperative of Responsibility*, 136; 140.

3

TECHNOLOGIES OF DESTRUCTION: THE SHADOW OF THE BOMB

Introduction: The Uniquely Destructive Character of Modern Technology

The critical conceptualisation and historicisation of the idea of technology laid out in the previous chapters accompanied a wider popularisation of the term 'technology' in the period following the Second World War. In both cases, this was related to material developments: new types of technology and innovations in existing technologies, as well as the increasing prevalence of technologies across all areas of public and private life. For the critics of technology, this continuous expansion of different technologies influenced their idea of 'technology' as embodying a certain set of traits; this idea of 'technology', in return, then shaped how they interpreted the continuing development of technologies. This was a discourse that was shaped by the reality of technology, in other words, not just by abstract theorisations. Alongside this, the political context – the rise, in the first half of the twentieth century, of a new form of politics in the shape of totalitarianism, as well as the post-war dominance of the two ideological opponents, and nuclear superpower states, of the USSR and United States – cast a deep shadow upon the critics' reflections on technology.

The increasing use of the term in the years following the Second World War had much to do with the invention and utilisation of increasingly destructive technologies of war, not least the astonishing power of atomic weapons, but also the automation of technologies of warfare. The critics of technology emphasised the intrinsic connection between technology and destruction: the total wars of the recent past and the feared atomic holocaust of the future arose out of this. 'Since violence', wrote Arendt, 'always needs *implements*... the revolution of technology, a revolution in tool-making, was especially marked in warfare.'[1] Technology's character as a destructive force extended beyond the obvious threats, however. More insidious forms of destruction loomed: the pollution of the planet by nuclear waste products (for example), or the destruction of the human psyche, or of social and ethical norms, that resulted from technology's integration into the everyday of contemporary modernity. The Cold War was the broad political context for these claims and directly influenced the idea of technology that emerged within the work of these thinkers. But their experiences of the Second World War and the period preceding it were also important to their analysis. Both total war and totalitarianism are enabled by modern technology, they argued, and furthermore, technology itself presses in that direction. It is clear that, for these authors, the Cold War, following on from totalitarianism and total war, in many ways represents its continuation.

The critics of technology agree that modern technology is uniquely destructive. While all technologies have a destructive quality (which need not in itself constitute a problem for politics), contemporary technology is excessively and extraordinarily destructive in ways which are historically disruptive. Modern technologies of war are distinct from earlier technologies of war because, where earlier technologies were limited in potential and use, contemporary technologies of war are total in scope and ambition. Technology's destructive capacity cannot be clearly separated from its use. It is not simply the case, they argue, that bad decisions lead to bad use of technology, and that negative consequences can be overcome by judicious and careful utilisation of these new resources. The very existence of these technologies must be considered destructive, not neutral, and the technologised societies that produce and advance such technologies equally self-destructive. The political outcome, these thinkers argue, is that such technologies realise and reflect an impulse to totalitarianism. They consider modern technological society to lead almost irrevocably towards a politics increasingly totalitarian in nature.

For each of these theorists, the emergence of twentieth-century totalitarianism played a central role in their political thought. Although Arendt wrote most extensively and influentially on the topic, ideas and fears of totalitarianism permeate the work of each of these thinkers. Largely, their concepts of totalitarianism seem similar to one another: that it entails the elimination of freedom, public and private; that it creates and utilises, through propagandistic means, and often very effectively, ideological justifications for its actions that mask or repudiate reality; that it is a highly bureaucratised state, albeit often 'irrational' in more traditional political or economic terms; that it describes a state in which plurality or diversity in politics or civil society are eradicated, and in which all parts of the state and society are oriented towards the central goals of totalitarian domination. It is thus essentially destructive: of its enemies or critics, of freedom, of politics as it has been traditionally understood. There are differences, though. Mumford is a notable outlier, offering a much more generalised definition of totalitarianism when he identifies totalitarianism in the ancient world as well as (in a reconstructed form) in the modern world. Totalitarianism, for him, and unlike the other theorists here, is not only a contemporary phenomenon – although, in modernity and the ancient world, he associated it with their respective forms of authoritarian 'megamachine'. There are differences of emphasis on the influence of political ideology in the development of totalitarianism, or the significance of experience: Arendt insists that totalitarianism can only be said to have come into the world with its realisation in Nazi Germany and Stalin's Soviet Union; the other theorists do not necessarily make the distinction between idea and experience in the same way, simply highlighting the gradual development of totalitarian-type politics in the modern world.[2] It is also significant that Arendt does not, in her 1951 *The Origins of Totalitarianism*, highlight the development of technology as one of the more important aspects of the rise of totalitarianism, although she identifies similarities between, for instance, totalitarianism and nuclear warfare. It is only later on that she develops the connection, most importantly in her 1958 *The Human Condition*, by which time she comes to associate totalitarianism much more closely with contemporary technology, and certainly as the space in which we are now at greatest threat *from* totalitarian trends.

There are several ways in which modern technology's uniquely destructive character is recognised in the work of the technological critics. It can be seen in contemporary techniques and technologies of warfare and terror, both 'conventional' and nuclear, although little was conventional about gas chambers, or rocket bombs, prior to the 1940s. These war machines represent the pinnacle of technological development. 'Nothing equals the perfection of our war machine,' wrote Ellul. 'Warships and warplanes are vastly more perfect than their counterparts in civilian life.'[3] Mumford argued that the modern megamachine – a unified system composed of both human and mechanical parts, and which comprised an assemblage *for* domination – 'could be brought about only under the fusion heat of war'.[4] The technologies of total war, or total annihilation, are latterly joined by the extraordinary destructive capacity of atomic and hydrogen weapons. But destructive capacity extends beyond even this. Technologies of destruction need not be immediately or deliberately destructive to warrant the label. The destruction of environment that inevitably occurs through technology's accumulation of waste or its various side effects is also highlighted by many of these authors, with Jonas particularly centring his critique of modern technology upon the inadvertent environmental destruction that results from its adoption. Finally, writers including Marcuse and Mumford claim that modern technology normalises and promotes destructive and aggressive behaviour in humans and human society.

These different forms of destructive capacity are linked to the development of specifically modern technology *beyond* the destructive powers of earlier technologies of warfare: the 'total' character of modern technologies of war makes irrational the act of war itself; the scope of environmental degradation and catastrophe goes beyond anything earlier technologies might have been responsible for; and while wars have inevitably stoked and utilised the aggression of its soldiers, the extent of technology's drive to aggression reaches into every corner of peacetime society. In short, technology's uniquely destructive capacity is the capacity to totally destroy, a distinction encapsulated by the power of the atom and later hydrogen bomb, but reaching far beyond that emblematic Cold War weapon. Modern technology's even more terrible quality is the apparent inevitability of this destructive potential.

Technologies of War

The wars of the twentieth century had seen devastating innovation in technique and technology. Had Hegel seen the present age, wrote Adorno, 'Hitler's robot-bombs' would no doubt have been seen as 'one of the selected empirical facts by which the state of the world-spirit manifests itself directly in symbols. Like Fascism itself, the robots career without a subject. Like it they combine utmost technical perfection with total blindness. And like it they arouse mortal terror and are wholly futile.'[5] The invention and widespread use of labour and death camps represents, Adorno argues, 'absolute integration . . . absolute negativity': the levelling off of individuality through the literal extermination of individuals. [6] Beginning in Auschwitz, wrote Anders, human beings were transformed into mere 'raw material', manufactured into corpses.[7] The Nazis used 'efficient modern technique . . . for their death factories' agreed Arendt.[8] Furthermore, she wrote, 'this terror finds its equivalent in total war, which is not satisfied with destroying strategically important military targets, but sets out to destroy – because it now technologically can seek to destroy – the entire world that has arisen between human beings.'[9] Wars in the twentieth century, she concludes, 'are not "storms of steel" (Jünger) that cleanse the political air, nor are they the "continuation of politics by other means" (Clausewitz); they are monstrous catastrophes that can transform the world into a desert and the earth into lifeless matter.'[10] With the atom bomb, humanity's destructive power reached extraordinary new heights. 'The atomic bomb has altered overnight the entire international picture,' wrote Mumford, two days after the attack on Hiroshima. 'The presence of this new source of energy makes every other form of military power, and every claim based on it, negligible.'[11] With the 'fact' of the atom bomb, Ellul argued, 'human beings divest themselves of their true superiority, and those who claimed to dominate things as well as the world now make themselves the slaves of facts, in a way that no dictatorship of the mind has ever dared hope was possible'.[12]

Heidegger, Anders and Arendt all wrote specifically on the existential transformation brought about by the birth of nuclear weapons. For Heidegger, the atom bomb's cataclysmic quality is explicitly representative of modern technology's intrinsic nature. 'What is dangerous is not technology,' Heidegger claims. 'The threat to man does not come in the first instance of the potentially lethal machines and apparatus of technology.'[13] Instead, he argues, we should look at how modern technology orders the world in such a way that blocks 'the shining-forth and holding-sway of truth', denying the possibility of a more 'original' understanding.[14] Technology, 'concealed' in the technological, is dangerous because it blocks access to a more authentic understanding of being itself.

This highly abstracted perspective on the meaning of the atom bomb contrasts with the work of those such as Arendt or Anders, who are all too aware of the tangible horrors and appalling human impact of the bomb. Nonetheless, while Heidegger's approach is more extreme in its emphasis on 'technology' (in the abstract) above the 'technological' (particular technologies), the idea that the atomic bomb represents the fulfilment of the development of a type of technology that is much broader in nature than *just* the bomb, or even technologies of war, is widespread among these critics of technology.

Anders wrote extensively on the dangers of nuclear weapons and spent decades campaigning for nuclear abolition when he returned to Europe after the war. Unlike Heidegger, Anders did not dismiss the deadliness or the physically destructive power of the atom bomb, but neither did he believe that its impact upon the modern world could be understood in the purely material terms of such destructive power. With the atom bomb, a shift in the nature of contemporary reality had taken place. In his foreword to the second volume of *The Obsolescence of Man*, Anders wrote:

> the world we live in today, and which surrounds us, is a technical world – which goes so far that we are no longer allowed to say that in our historical situation technology is only one among other things, but rather have to say: history now takes place within the state of the world known as 'technology' . . . [specifically] it is wrong to claim that the atom bomb exists within our political situation, it is instead correct to say that politics now finds its place within the atomic reality.[15]

The bomb initiated what Anders calls 'the third industrial revolution', a revolution entailing the 'spectacular' transformation of humanity.[16] For the very first time, humanity is in the position to 'produce its own destruction' through the atomic bomb, he writes.[17] Furthermore, and even more disturbingly, Anders argues that 'it belongs to the essence of our technical existence, that we not only cannot or may not *not-produce* what we can produce, but also . . . we cannot or may not *not-use* what has been produced' (emphasis added).[18] We are thus compelled to act in certain ways. As such, the third industrial era initiated an epoch in which we 'constantly pursue the production of our own destruction (what we do not know is only the exact point in time at which this will occur)'.[19] The continuation of history is an assumption no longer permitted, and this, he argues, is the metaphysical transformation brought about by the physical reality of atomic weapons.[20] Because he believes it is of the nature of a technologised society to produce and to consume in ever greater quantities, and that technology presses us towards this, the atomic bomb can never be considered a 'neutral' weapon: it can only ever mean inevitable civilisational or global destruction. The nuclear epoch is, Anders concludes, 'whether it ends or continues, the last, because the danger we have put ourselves in through our spectacular product, and which has now become the final mark of Cain of our existence, can never end – even through the end itself'.[21]

Arendt, as Yaqoob explains, also reflected aspects of, but 'was not simply reiterating Heidegger's treatment of technology . . . Rather than treating science and technology in terms of unfolding essences, Arendt sought to stress their contingent development.'[22] Like Anders, Arendt stresses the transformational destructive capacity of the atom bomb, and the qualitative change in modernity that this quantitative advancement of military power had brought about. In all ages, she writes, societies have engaged in warfare and utilised tools of destruction and violence. Normally, however, the ability to destroy and produce stand in some approximation of balance, she writes; the possibility of rebuilding has always been present. Harnessing, or unleashing, nuclear energy changes all this:

> as long as the abilities to produce and destroy stand in balance, everything proceeds more or less as it always has . . . A change in all this was possible only with the discovery of atomic energy, or better, with the invention of a technology driven by the processes of nuclear energy, for it is not natural processes that are unleashed here. Instead, processes that do not occur naturally on earth are brought to earth to produce a world or destroy it.[23]

Nuclear weapons mark a qualitative shift in the nature of destructive technology not only because they disrupt the balance of productive and destructive capacity, but because they initiate a wholly new form of technological process: the unleashing of natural, 'universal' process into the earthly domain. For Arendt, nature represents a force that humanity has struggled against throughout existence: we are organic beings, and thus rely upon nature's resources and the environment it provides, yet at the same time nature is a sphere of fundamental violence, force and necessity. Nature cannot be a space for freedom,

Arendt insists. Our interactions with nature via labour, for instance, are defined by her as being strictly necessary acts. Only by creating artificial spaces *outside* necessity – political communities, such as the city, classically conceived – have humans been able to carve out a space for freedom. Politics and freedom stand against nature. Through unleashing even more powerful natural forces, such as those that take place inside stars, into the world, the precarious balance between natural forces and the small, fragile spaces of human freedom that exist within the artifice of constitution and legislation is also unsettled. Like Anders and Heidegger, Arendt identifies inevitability in nuclear devastation, and this fact shaped her political ideas from the 1950s onwards. 'Arendt's suggestion that "now-familiar debates" needed to be superseded must be understood in the specific context of nuclear power', argues Yaqoob.[24]

Technology's Destructive Dynamic

Catastrophic as the new technologies of war might be, the destructive capacity and quality of modern technology goes beyond weapons of total war and even the apocalyptic threat of nuclear war. Technology destroys in unintentional ways as well as through design. This is observed by these thinkers in the side effects of both destructive *and* notionally productive technologies: the effects of industrialised agriculture as well as the broader consequences of producing and setting off an atomic warhead. Technology's ill effects are widespread, ranging from its attacks upon the natural environment to human nature itself.

'Man is about to hurl himself upon the entire earth and its atmosphere,' wrote Heidegger.[25] Atomic explosions, Ellul insists, 'are not the real problem. The real problem continues to be that of the disposal of the ceaselessly accumulating waste materials.'[26] Jonas, who, of these theorists, most extensively considered the environmental effects of technology, wrote that, in the latter part of his career (that is, from the 1960s onwards), he experienced:

> [a] growing realization of the inherent dangers of technology as such – not of its sudden but of its slow perils, not of its short-term, but of its long-term threats, not of its malevolent abuses which, with some watchfulness, which one can hope to control, but of its most benevolent and legitimate uses which are the very stuff of its active possession.[27]

Atomic holocaust is indeed possible, even plausible, wrote Jonas, but much more so is rising production, consumption and population growth. 'The net total of these threats is the overtaxing of nature, environmental and (perhaps) human as well,' he argues.[28] His main fear is not the 'predictable' and highly visible atom bomb, he explains, but rather 'the apocalypse threatening from the nature of the unintended dynamics of technical civilization as such . . . whereto it drifts willy-nilly and with exponential acceleration: the apocalypse of the

TECHNOLOGIES OF DESTRUCTION

"too much," with exhaustion, pollution, desolation of the planet.'[29] Unlike possibly, even plausibly, avoidable nuclear destruction, the apocalypse of the 'too much' is 'almost bound to come by the logic of present trends that positively forge ahead toward it'.[30]

Both Mumford and Ellul agree, pointing out that the environmental impact of technology is already evident. Modern technology has been completely 'divorce[d] from ecological moderations and human norms,' writes Mumford. 'Who can doubt that the destructions and the massacres, the environmental depletions and the human degradations that have become prevalent during the last half century have been in direct proportion to the dynamism, power, speed, and instantaneous control [of megatechnics]?'[31] We *know* the damage we do, notes Ellul, but this knowledge makes little difference:

> We know the implications of pollution, but we go on calmly polluting the air, the rivers, and the ocean. We know men are going mad from living in huge conglomerations, but we, like automatons, go on building them. We know the dangers of pesticides and chemical fertilisers, but we continue to use them in increasingly massive doses. We know all this, but we are like the masochist who knows others have put a little arsenic in each bowl of soup he drinks, but who goes on drinking it day after day, as though impelled by a force he cannot resist. Our speed is constantly increasing, and it does not matter whither we are going.[32]

Marcuse and Mumford also argue that technology is causing human nature to change: to become more oriented towards destructiveness by unleashing its aggressive instincts. In two essays, written in 1966 and 1967, Marcuse argues that technology's incorporation into modern society has had a transformational effect on the aggressiveness of those living within it. '*Technological aggression and satisfaction*' is a distinctly new phenomenon, he argues.[33] The aggressive act itself is no longer physically undertaken by the human agent, but rather by much more powerful machines, which carry out the act automatically. 'This means that the energy, the power activated and consummated is the mechanical, electrical, or nuclear energy of "things" rather than the instinctual energy of a human being,' Marcuse writes. He suggests that the 'delegation' of aggression to intermediaries causes the 'instinctual satisfaction of the human person' to be '"interrupted," reduced, frustrated, "super-sublimated"'.[34] Such frustration, he concludes, 'makes for repetition and escalation: increasing violence, speed, enlarged scope', which along with a weakened sense of guilt or responsibility causes the aggressive behaviour to accelerate.[35]

> The fact that aggression and destruction are carried out by a thing – a mechanism, an automated device – rather than by a person, impairs the

> satisfaction of the aggressive instinct, and this frustration prompts repetition and escalation of aggression . . . The result is brutalization on a massive scale.[36]

'The automaton was not born alone,' wrote Mumford, echoing Marcuse's sentiment. 'The automaton has been accompanied, we can now see, by a twin, a dark shadow-self . . . aggressively destructive, even homicidal, reasserting the dammed-up forces of life in crazy or criminal acts.' In this new man, Mumford argues 'the reflexes and blind instincts [are] in command . . . destroy[ing] those higher attributes of man whose gifts of love, mutuality, rationality, imagination, and constructive aptitude have enlarged all the possibilities of life'.[37]

The atom bomb is a highly visible representation of the destructive essence of technological modernity for all these thinkers, but the destructive impact of technology reaches far beyond the immediate consequences even of nuclear warfare. Modern technology is destructive to an extent that is transformational for human societies, and it is destructive beyond just its technologies of warfare. Its innately destructive nature has several interlinking components. In part, the impact of technology is intentional in the sphere of war and conflict, but their unintentionally destructive qualities are arguably greater: on the earth's environment, on human 'worldliness', on the balance between nature and artifice, or on human nature itself. This is not a world in which technology merely plays a part, but a world which is primarily framed *by* technology, and from which there is seemingly no possibility of return. Because technology appears to possess a self-propelling or autonomous development for these thinkers, the unintended, unknown consequences of technological advancement will always be greater than what is planned. This results in uncontrollable cycles of technological acceleration and intensification leading ultimately to a totalisation of destruction: this is an apocalyptic vision.

From the Second World War to the Cold War

The nuclear attacks on Hiroshima and Nagasaki opened a new and dreadful chapter in the history of warfare and of technology. Yet, while this was something objectively novel in human history, it was not unexpected, argue our thinkers, but constituted the culmination of a trend of technological development that began long before. The events of the bombings mark a moment of transformation, the flowering of destructive technology's potential, but earlier events had made this seemingly almost inevitable. 'The atomic bomb and its explosion', wrote Heidegger, is merely the 'latest emission' of what 'for a long time now *has* already arrived'.[38] For Heidegger, the atom bomb is merely the physical realisation of modern science and its destruction of a traditional way of understanding and ordering the world: 'the crudest of all crude confirmations of the annihilation of things that occurred long ago: confirmation that the thing as thing remains nullified'.[39] Yet the atomic explosions are also the origin of the 'modern world', wrote Arendt; a new political era, channelling nature's 'elementary power, into the world itself'.[40] Mumford wrote of how, until the 1940s, 'it was still possible', although essentially misguided, 'to regard the continuation and acceleration of modern technology as, on the whole, favorable to human development'.[41] Such a thought would become impossible with the emergence of nuclear weapons, he asserts. Behind this new innovation, however, was the older dynamic of the military machine: 'repeating cycles of extermination, destruction, and self-extinction'.[42] Now, as before, the 'two poles of civilization . . . are mechanically organized work and mechanically organized destruction and extermination'.[43] When atom bombs were succeeded by the exponentially more powerful hydrogen bombs, the critics of technology saw only a predictable acceleration of the destructive potential in atomic weapons, wholly in accordance with the general tendencies of technology towards greater efficiency and efficacy.

Given that Anders, Arendt, Adorno, Horkheimer, Jonas and Marcuse were all German-Jewish exiles whose lives were profoundly affected by the rise of Nazism, and who were deeply disturbed, personally and intellectually, by the Holocaust and the actions of the totalitarian regimes, it is far from surprising that this fundamental context influenced their political reading of the new technologies of destruction. This clearly the case for all of these thinkers. Adorno's 'most pessimistic conceptualization . . . of technology and scientific rationality' was rooted in 'the rise to power of the Nazis and the industrialized horrors of the holocaust [which] played no small part in the formulation of his ideas', writes Thomas Sheehan.[44] Arendt wrote of the unprecedented collapse caused by the total wars of the twentieth century: 'We no longer hope for an eventual restoration of the old world order with all its traditions . . . Never has our future been more unpredictable.'[45] For Jonas, the Second World War revealed an apocalyptic state of the world and necessitated a complete re-evaluation of being.[46] The absolute destructiveness of the era had a formative impact on how they understood modernity.

The context of the Second World War and the years preceding it were also pivotal for Ellul, Mumford and Heidegger, although their experiences were quite different from those exiled. However, as was the case for many Europeans, Ellul's political awareness developed in the context of the rise of European fascism and his experience of the war.[47] Mumford, in 1951, wrote of how society is riddled with the agents of 'disintegration and barbarism', those such as 'the connoisseur of violence who has devaluated everything about life except the instruments for defacing it, the inventors of the extermination camp, the agents and potential practitioners of random violence who devise H-bombs and biological instruments of genocide'.[48]

In the case of Heidegger, although he clearly identifies – in his post-war work – a long-standing tendency towards annihilation in modernity, the connections that he draws between this and technology itself have a more complex development over his career. Jeffrey Herf argues that, in the 1930s, Heidegger held out 'hope that Germany would be the country to achieve a fusion of technology and soul'. As such, Herf claims, he 'made a tenuous peace with both Nazism and technology, whatever his postwar retrospectives on technological dehumanisation may have been'.[49] Vincent Blok, on the other hand, notes that 'Heidegger remarks that because of his reading of Jünger, he had seen very early on what the dreadful events of the Second World War would confirm much later: "The universal dominion of the will to power within planetary history."'[50] Ihde explains that, while 'the hyper-masculinist, nationalistic take on technologies does not appear with the postwar 1950s writing', it does not disappear; 'rather, it *shifts*'.[51] Heidegger comes to associate *technê*, 'a poetic production process which he claims from the Greeks', with both pre-modern technologies and works of art.[52] These various positions are not necessarily at odds with one another. It is obviously true that Heidegger threw in his lot with the Nazis, and furthermore, after the war, refused to publicly apologise for his actions. But it is not so clear how far Heidegger was persuaded that Nazism could provide the solution to the problem of technology in modernity.

By the late 1940s, whatever their earlier ideas of technology had been, and despite the different experiences of these thinkers during the 1930s and 1940s, their readings of what the war had meant came to intersect with one another. That is, that total destruction and totalitarian politics was implicit in the development of modernity, and this was neither a specifically German phenomenon nor a phenomenon whose influence was exclusive to those states widely recognised as totalitarian. The invention and use of extraordinary technologies of destruction cut across liberal and authoritarian states in the war; it was the United States, of course, that unleashed the atom bomb upon the citizens and cities of its foe. The trauma and the tragedy of the war revealed the collapse of civilisation in modernity, where the only winner of the war was the technology and technique which ferociously advanced into the Cold War era. In his Bremen lectures, the immediate precursor to 'The Question Concerning Technology', Heidegger makes the claim that 'Agriculture is now a mechanized food industry, in essence the same as the production of corpses in the gas chambers and extermination camps, the same as the blockading and starving of countries, the same as the production of hydrogen bombs.'[53] From one perspective, this is astonishingly lacking in empathy for the victims of the chambers and camps, and equally the victims of nuclear war, because of its obliviousness to the clear moral distinctions between these various situations. Yet, it is also the case, perhaps more surprisingly, that most of our other critics of technology would largely agree with the sentiment Heidegger expresses here: that the murderous realities of the war era relied upon the mechanisation and technologisation of society at every level, and that this was embedded even more deeply in society through the war itself.

THE POLITICAL CONSEQUENCES OF A TECHNOLOGY OF TOTAL DESTRUCTION: THE COMPREHENSIVELY TOTALITARIAN NATURE OF MODERN POLITICS AND SOCIETY

This vision of an inherently destructive, catastrophic technology plays a key role in the understanding of totalitarianism in these theorists' work. Technology impacts modernity in the widest possible sense, but from a political point of view these thinkers are united in their claim that modern technology is totalitarian in nature, and its growth supports and tends to the furtherance of totalitarian politics. It is not the case that technology is the only influence on the development of totalitarianism: multiple political, social, cultural, economic (etc.) aspects shaped its emergence. Nor is it simply the case, as should already be quite clear, that technology only provides the technical means for totalitarian politics (although it is true that modern technological advances are required for totalitarian rule). Rather, technology, in the form it appears in modernity, is itself 'totalitarian' in nature and thus drives totalitarian politics in technologised societies. 'Technique cannot be otherwise than totalitarian,' writes Ellul. 'Totalitarianism extends to whatever touches it, even things which seem, at first sight, very remote from it . . . Technique can leave nothing untouched.'[54] Most of the time, Anders argues, totalitarianism is considered as a political form, yet he insists this is not a true representation of it. Instead, he writes, 'the principle of totalitarianism is a technical principle'.[55]

> I put forward the thesis that the tendency towards totalitarianism belongs to the nature of the machine and originally stems from the field of technology; that the tendency, inherent in every machine as such, to overwhelm the world, to parasitically exploit it, to coalesce with other machines, and to function with them as parts within a single total machine – I argue that this tendency constitutes the basic fact; and that political totalitarianism, however appalling, is only an effect and variant of this basic technological fact.[56]

Machines possess, Anders argues, a 'will to power . . . the expansionist urge of the machines is insatiable'.[57]

Arendt ties together totalitarianism and the atom bomb as the two 'fundamental experiences of our age'.[58] Both force us to question what the meaning of politics might be, she argues, indeed whether politics still has any meaning today. Totalitarianism fundamentally threatens freedom and thus the political world in which we live. The atomic bomb threatens freedom but also life itself, even the possibility of life. If this is politics today – totalitarianism plus technologies of nuclear destruction – such a politics seems only to bring inevitable disaster. In such a world, it can seem, Arendt writes, that our only hope is to attempt to do away with politics altogether, in the hope that such devastation might be avoided (a utopian hope that Arendt has little truck with).[59] As Jonathan Schell highlights, totalitarianism and nuclear war both appear in a fundamentally similar light in Arendt's work, as the elimination of all possibility for a common world.[60] While Arendt's analysis of totalitarianism – its character, the way in which it functions, and the history of its emergence in the Western world – is complex, perhaps the most important and characteristic feature of totalitarianism is its absolute opposition to human spontaneity or freedom. The concentration camps were therefore the definitive organisational structure of the totalitarian regimes, she argued, because, in them, the total elimination of spontaneity was realised. And yet, it is also the case that Arendt was doubtful as to the possibility of the absolute elimination of spontaneity. As long as humans lived, there remained a chance for free action. The ultimate end of totalitarianism was not simply control over humans, but the elimination of all potential spontaneity via the elimination of all human beings. The inherent destructiveness of totalitarianism is taken forward in her work in the context of her developing interest in technology.

The supposed 'neutrality' of technique, argues Marcuse, 'is in itself a political concept'. While technics 'projects instrumentality as a means of freeing man from toil and anxiety', it has developed into pure instrumentality, absent of this original final cause. 'Hence, pure instrumentality, without finality, has become a universal means of domination . . . mastery of nature insofar as it is a hostile, violent, and destructive force; mastery of man to the extent that he is a part of that nature.'[61] As Dana Villa explains, here technology 'is not simply an ideology masking domination, but a form of domination itself'. Marcuse 'insists on the deformations of a technological civilization, and not just the distortions of a technocratic ideology'.[62]

The ways in which totalitarianism utilises technology, or the ways in which totalitarian politics might effectively use modern technologies of war to entrench its domination are self-evident. The more interesting set of claims made by the critics of technology surround the way in which technology itself drives totalitarianism, or even impels totalitarian politics in modernity. A society that is technological *inevitably* becomes politically totalitarian, because of the totalising, destructive nature of modern technology: 'the rationale of domination itself'.[63]

The Totalitarian Character of Modern Technology

The two experiences of 'totalitarianism and the atomic bomb', writes Arendt, 'ignite the question about the meaning of politics in our time. They are the fundamental experiences of our age, and if we ignore them, it is as if we never lived in the world that is our world.'[64] Technology destroys the world as we have known it, and potentially the world as it is viable for human life. Jonas speaks of the 'apocalyptic pregnancy of our actions', while, Mumford argues, the ultimate consequence of Galileo's 'mechanical world picture' was an environment suited only to machines, not man.[65] Today's complete politicisation of society 'is merely an offshoot of the machinational essence of beings', Heidegger wrote, and 'machination retains genuine power all the more securely, the more exclusively the execution of political power considers itself the be-all and end-all'.[66]

Along with the extraordinary increase in power that modern technology offers is the escape of that power from the control of human hands. 'Men pay for the increase of their power with alienation from that over which they exercise their power,' warn Adorno and Horkheimer.[67] In all technologically developed states, Arendt argues, the relationship between power (deriving from humans acting together in the political sphere) and violence means that power is, uniquely in human history, completely overwhelmed by the latter.[68] With 'the accumulation of techniques in the hands of the state', writes Ellul, the state becomes totalitarian. Interconnected techniques come to form a system that 'tightly encloses all our activities', he claims. 'When the state takes hold of the single thread of this network of techniques, little by little it draws to itself all the matter and the method, whether or not it consciously wills to do so.'[69]

It is not only that technology puts absolute control in the hands of a minority, but that ultimately it takes power out of human control altogether, according to these theorists. Technology is totalitarian because its dynamic evades human control, and therefore comes to be a force of necessity, not freedom. Because technology follows its own principles of action, it therefore acts against human autonomy or control, and is neither a neutral force nor a force for freedom. The technological principle is 'efficient ordering' (Ellul) or 'overwhelming efficiency' (Marcuse).[70] 'It is *necessity* which characterizes the technical universe,' Ellul insists. The outcome, argues Marcuse, is that 'concentration camps, mass exterminations, world wars, and atom bombs are no "relapse into barbarism," but the unrepressed implementation of the achievements of modern science, technology, and domination'.[71] The existence of technique 'forces political choice', writes Ellul; likewise, Marcuse explains, the technological apparatus 'functions, not as the sum-total of mere instruments, but rather as a system which determines *a priori* the product of the apparatus'.[72] Because of this, Marcuse concludes, 'technological rationality has become political rationality'.[73] Anders makes the same point, when he writes that 'the tendency towards totalitarianism is part of the essence of the machine . . . this tendency represents the fundamental fact and . . . political totalitarianism, as horrible as it is, only represents an effect and variant of this fundamental technological fact'.[74]

Technological rationality, however, is wholly at odds with rationality understood in any humanistic or traditional manner. Technology, in fact, from a human perspective, is wholly irrational, because the efficiency that technology aims towards is the extension of technology's own efficacy; technology is not directed towards (human) ends. The means of technology has become its own end. As such, Ellul writes, 'today everything has become means. The end no longer exists . . . [and] the problem of means today is that they are totalitarian.'[75]

This is true across every modern civilisation in the broadest sense. Thus, writes Heidegger:

> machination eludes all explanations that are based on human activities; – qua being, it penetrates all humanity and the remainder of the 'human' world which is hollowing itself out and is driven into decisionlessness . . . The measurelessness of machination demands no human presumption; it demands only the ungrasped and unknown detachment from every essential decision.[76]

The formalisation of reason, Adorno and Horkheimer explain, 'caused all goals to lose, as delusion, any claim to necessity and objectivity. Magic is transferred to mere activity, to means.'[77]

Arguably, however, total war is the ultimate reflection of this irrationality for most of our critics of technology. 'If war signifies "total war," i.e., the one deriving from the unfettered machination of beings as such,' writes Heidegger, 'then it becomes a *transformation* of "politics" and a *revelation* of the fact that "politics" itself has become merely the executor of unmastered metaphysical decisions, an executor that is no longer in charge of itself.'[78] Although modern robot bombs possess 'utmost technical perfection', Adorno observes, they combine it with absolute blindness as to what they are doing.[79] Through atomic warfare, a war of total annihilation, war can no longer be a rational means of politics, Arendt claims. And although 'total war was first proclaimed by nations under totalitarian rule . . . in doing so they inevitably forced their own principle of action onto the nontotalitarian world. Once a principle of such vast scope enters the world, it is of course practically impossible to limit it.'[80]

Technology and the Cold War World

The centring of technology in their work, and the relationship that they identify between modern technology and totalitarianism, shapes the way these theorists understand the Cold War world. First, there is a recognition of a mode of globalisation taking place in modernity by which Western imperialism continues and expands via the means of technology and its effects. This is transforming the geopolitical context. Second, because technology is seen as something that is essentially identical in both the Western and communist worlds (as well as in fascist states), albeit realised or utilised in (perhaps temporarily) divergent ways, the driving impulse and direction of both are towards an increasingly totalitarian politics. Technology thus creates negative unity in the world: making less developed states more like technologically advanced nations, and unifying the opposing forces of Western liberal democracy with Soviet communism through an underlying shared reliance upon technology. 'Behind the democratic or totalitarian aspects there is the reality of the technical state, which pursues its course regardless of whatever exterior form it may take,' writes Ellul.[81]

'Mankind', as a unified entity, 'owes its existence not to the dreams of humanists nor to the reasoning of the philosophers and not even, at least not primarily, to political events, but almost exclusively to the technical development of the Western world,' writes Arendt.[82] 'Technology united the world.'[83] The unity of the world – that humanity possesses, for the first time in history, some commonality – is not necessarily a negative development. 'No event of any importance can remain a marginal accident in the history of any other . . . every man feels the shock of events which take place at the other side of the globe.'[84] Yet, for Arendt, this does not comprise politically meaningful unity because 'this common factual present is not based on a common past and does not in the least guarantee a common future'.[85] Technology might unify, but it can just as easily annihilate.

> At this moment the most potent symbol of the unity of mankind is the remote possibility that atomic weapons used by one country according to the political wisdom of a few might ultimately come to be the end of all human life on earth. The solidarity of human life in this respect is entirely negative.[86]

It is a very different kind of 'world' from the community created by the genuinely political societies Arendt describes in her work, in which citizens express their power by freely acting together and which are tied together by tradition, legislation and norms.

'Though for twenty years after the atom bomb was dropped only two modern military megamachines came into existence,' writes Mumford, 'those of the United States and Soviet Russia – they both carry the prospect . . . of drawing into their orbit every other national unit.'[87] 'Technique is constantly gaining ground, country by country and . . . its area of action is the whole world,' says Ellul. 'In all countries, whatever their degree of "civilization," there is a tendency to apply the same technical procedures.'[88] Through commerce and war, a technical invasion has taken place. The 'admiration and fear' of the vanquished caused them to adopt the machines which, argues Ellul 'came to replace their gods'.[89]

> Our Western civilization has imposed its mechanical and rational mold on the whole world, but it leads to a fatal deadlock. Disaster in all its forms has fallen upon the entire earth as never before. Totalitarian wars, dictatorial empires, administratively organized famines, complete moral breakdown in contexts both social (nation, family) and internal

(individual amorality), the fabulous increase in wealth that does not benefit the most destitute, the enslavement of almost all humanity under the domination of states or individuals (capitalism), the depersonalization of humanity as a whole and individually – all this is well known . . . The more progress we make, the more we prove ourselves incapable of ruling and directing the world that has issued from our hands.[90]

Totalitarian technology is the underlying context of the Cold War. While modern technology is, these thinkers agree, an originally Western phenomenon, the Second World War, as well as the Cold War, provided a stimulus to the development of technology's more political orientation and influence. It is interesting that perhaps the most traditionally 'liberal'-minded theorist of these critics of technology, Lewis Mumford, makes this argument very clearly. He outlines the 'reassemblage' of the dominating megamachine in modernity, with the final (or most recent) step being made by the Nazi and Soviet megamachines, which took technology and technocracy to its next stage of development. Despite their ultimate military failure, the Nazis achieved, during the war, 'a series of astounding military successes; and these brought about a similar recrudescence of the megamachine in Britain and the United States', Mumford wrote. 'By the curious dialectic of history, Hitler's enlargement and refurbishment of the megamachine gave rise to the conditions for creating those counterinstruments that would conquer it and temporarily wreck it.'[91] The Nazi megamachine did not repulse Western democracies, or discredit the notion of megamachine; it rather *inspired* the West to rebuild it.[92] In resisting fascism in the Second World War, Mumford argues, the Western democracies 'abandoned the moral standards and laws of war hitherto respected by "civilized" nations and copied the abominable fascist practice of indiscriminately exterminating civilian populations'. The precedent had thereby been set that 'justified' the use of the atom bomb. The carpet-bombing of German cities, for example, was denounced by Mumford as a symptom of America's decline into barbarism.[93] Such actions of total annihilation could only be described, Mumford writes, as 'totalitarian de-moralization'.[94]

Even where differences were conceded between the dictatorships and democracies of the world by the critics of technology, these were considered to be in practice temporary and partial. 'Dictatorships are conscious of technique and try to use it, whereas democracies are not, they are inhibited by their scruples,' yet more often than not, Ellul suggests, 'these are really nothing more than "smokescreens"'.[95] The traditional forces of restraint – morality, public opinion and so on – increasingly had been brought into the cause of technique, argues Ellul. And 'cold war', he argues, 'is as productive as hot war in forcing the democracies to imitate the dictatorships in the use of technique'.[96] Mumford, too, explicitly likens 'Nazism and Stalinism and Nixonism . . . all the variants of an oppressive totalitarianism'.[97]

This claim is at the very heart of Adorno and Horkheimer's *Dialectic of Enlightenment*. 'The specter of totalitarianism still hovered over the political landscape when *Dialectic of Enlightenment* was published in 1947,' writes Stephen Bronner. Nazism had not won the war, but substantial similarities were clearly visible in the Soviet Union, while the proto-imperialist, profoundly racist United States was far from beyond reproach. 'Max Horkheimer and Theodor Adorno saw the future appearing as present: and it was not the future that the Enlightenment had foretold.'[98] Yet the source of technological modernity was the Enlightenment, they argued. Adorno and Horkheimer were concerned with showing, Bronner explains, 'not merely the internal connection between totalitarianism and the Enlightenment, but how the germs of barbarism existed in western civilization from the beginning'.[99] For Adorno and Horkheimer, writes Sheldon Wolin, 'the brute facts of the twentieth century – totalitarianism, racism, death camps, genocide, cultural barbarism, and the celebration of military aggression – made it no longer possible to believe in the inevitability of progress or the neutral character of modern technology'.[100]

A Retreat from Technology?

Technologies of warfare were evidently pivotal to the nature of the Cold War, particularly the existence of the nuclear threat. The critics of technology explored here understood this. However, they emphasise that technology did not just shape this war, but rather, that technology had structured modernity and modern warfare itself. As such, they believed that this was a war in which *both* superpowers and their allies were in essential respects alike politically, because in both societies technology was existentially transformational along the same lines: it leaned towards totalitarianism. The globalising nature of technology also meant that not only those engaged in this war, but every nation in the world, would soon come to orient themselves around the same technological and thus political system.

This represents, all these theorists believe, a fair representation of existing trends in world politics, but it is also a pessimistic vision of world politics, one which seemingly dismisses ideological visions, material difference and socio-political structures as somewhat superficial, ultimately meaningless or powerless against an autonomous, totalising technology. Of course, it is not the case that there are *no* meaningful distinctions between, for instance, liberal Western constitutional democracies and the extreme totalitarian societies of earlier Nazi Germany or Stalinist Russia. Yet technology's totalising, dominating influence is a powerful force upon society, and one which, these theorists suggest, will not be easily reversed.

However, technology has not yet attained absolute power, these theorists would agree. The success of the Vietnamese opposition to American firepower during the Vietnam War, as Arendt points out, is testament to the limits of the destructiveness of technology against the sustained resistance of a people willing to stand together to defend their cause.[101] Even extraordinary technological power does not (yet) have the capacity to overcome such an enemy. On the other hand, to retreat to an earlier age of politics in the West, to simply refuse to engage in the technological dynamic, is also not a viable option: 'The distressing thing about the emergence within politics of the possibility of absolute physical annihilation is that it renders such a retreat totally impossible.'[102] We might say, Arendt writes, that 'a decisive change for the better can come about only through some sort of miracle'.[103] Anders claims that 'to believe that such totalitarianism could be curbed or prevented through purely political means is naïve. For ultimately the root of this totalitarianism lies in a technical fact.'[104] Only a response that dealt appropriately with technology would be capable of reshaping politics, he argues. Yet it is very unclear what such a response might be. It is very difficult to see how the dynamic of a destructive, totalitarian technology such as has been set out by these thinkers might in practice be reversed or opposed.

NOTES

1. Hannah Arendt, 'On Violence' [1969], in *Crises of the Republic* (Harmondsworth: Penguin, 1973), 106.
2. This is a very significant distinction, particularly for Arendt whose methodology in *Origins*, of tracing 'subterranean currents' in Western history that led to totalitarianism and the *overturning* of traditional politics, stresses the radical and not gradualist character of the birth of totalitarianism.
3. Jacques Ellul, *The Technological Society*, trans. John Wilkinson (New York: Random House, 1964), 16.
4. Lewis Mumford, *The Myth of the Machine: The Pentagon of Power* (New York: Harcourt Brace Jovanovich, 1970), 236.
5. Theodor Adorno, *Minima Moralia: Reflections from Damaged Life* [1951], trans. Edmund Jephcott (London: NLRB, 1974), 55.
6. Theodor Adorno, *Negative Dialectics* [1966] (New York: Seabury Press, 1973), 362.
7. Günther Anders, *Die Antiquiertheit des Menschen*, vol. 2: *Über die Zerstörung des Lebens im Zeitalter der dritten industriellen Revolution* (Munich: Verlag C. H. Beck, 2018), 23.
8. Hannah Arendt, 'The Image of Hell' [1946], in *Essays in Understanding 1930–1954: Formation, Exile, and Totalitarianism* (New York: Schocken Books, 2005), 204–5.
9. Hannah Arendt, 'Introduction *into* Politics', in *The Promise of Politics* (New York: Schocken Books, 2005), 162.
10. Ibid., 191.
11. Lewis Mumford, *My Works and Days: A Personal Chronicle* (New York and London: Harcourt Brace Jovanovich, 1979), 438.
12. Jacques Ellul, *The Presence of the Kingdom*, trans. Olive Wyon (Philadelphia: Westminster Press, 1951), 23.

TECHNOLOGIES OF DESTRUCTION

13. Martin Heidegger, 'The Question Concerning Technology' [1949], in *The Question Concerning Technology and Other Essays*, trans. William Lovitt (New York: Harper & Row, 1977), 28.

14. Ibid.

15. Anders, *Die Antiquiertheit des Menschen*, vol. 2, 9; 481fn.

16. Ibid., 20.

17. Ibid.

18. Ibid.

19. Ibid.

20. Ibid., 21.

21. Ibid.

22. Waseem Yaqoob, 'The Archimedean Point: Science and Technology in the Thought of Hannah Arendt, 1951–1963', *Journal of European Studies* 44:3 (2014), 205.

23. Arendt, 'Introduction *into* Politics', 157.

24. Yaqoob, 'The Archimedean Point', 206.

25. Martin Heidegger, 'Anaximander's Saying' [1946], in *Off the Beaten Track*, trans. Julian Young and Kenneth Haynes (Cambridge: Cambridge University Press, 2002), 280–1.

26. Ellul, *The Presence of the Kingdom*, 109.

27. Hans Jonas, 'Introduction', in *Philosophical Essays* (New York and Dresden: Atropos Press, 2010), xvii.

28. Hans Jonas, *The Imperative of Responsibility* (Chicago: University of Chicago Press, 1984), 19.

29. Ibid., 202–3.

30. Ibid.

31. Mumford, *The Myth of the Machine: The Pentagon of Power*, 349–50.

32. Jacques Ellul, *The Betrayal of the West*, trans. Matthew O'Connell (New York: Seabury, 1978), 197.

33. Herbert Marcuse, 'Aggressiveness in Advanced Industrial Society' [1967], in *Negations: Essays in Critical Theory*, trans. Jeremy J. Shapiro (London: Free Association, 1988), 263.

34. Ibid., 264.

35. Ibid.

36. Herbert Marcuse, 'The Inner Logic of American Policy in Vietnam' [1966], in *Teach-Ins, USA: Reports, Opinions, Documents*, ed. Louis Menashe and Ronald Radosh (New York: Praeger, 1967), 65–6.

37. Mumford, *The Myth of the Machine: The Pentagon of Power*, 193.

38. Martin Heidegger, 'Insight into That Which Is' [1949], in *The Bremen and Freiburg Lectures*, trans. Andrew J. Mitchell (Bloomington and Indianapolis: Indiana University Press, 2012), 4.

39. Ibid., 5.

40. Hannah Arendt, *The Human Condition* [1958] (Chicago: University of Chicago Press, 1998), 149.

41. Mumford, *The Myth of the Machine: The Pentagon of Power*, 236–7.

42. Mumford, *The Myth of the Machine: Technics and Human Development* (New York: Harcourt, Brace & World, 1967), 211.

CATASTROPHIC TECHNOLOGY IN COLD WAR POLITICAL THOUGHT

43. Ibid., 221.
44. Stephen Sheehan, 'The Nature of Technology: Changing Concepts of Technology in the Early Twentieth-Century', *Icon* 11 (2005), 8–9.
45. Hannah Arendt, *The Origins of Totalitarianism* [1951] (Orlando: Harcourt, 1968), vii.
46. Jonas, 'Introduction', in *Philosophical Essays* (New York and Dresden: Atropos Press, 1980), xiii.
47. David Lovekin, *Technique, Discourse and Consciousness: An Introduction to the Philosophy of Jacques Ellul* (Bethlehem: Lehigh University Press, 1991), 133–4.
48. Lewis Mumford, *The Conduct of Life* (New York: Harcourt, Brace & Co., 1951), 222.
49. Jeffrey Herf, *Reactionary Modernism: Technology, Culture, and Politics in Weimar and the Third Reich* (Cambridge: Cambridge University Press, 1984), 44.
50. Vincent Blok, *Ernst Jünger's Philosophy of Technology* (New York and London: Routledge, 2017), 53.
51. Don Ihde, *Heidegger's Technology: Postphenomenological Perspectives* (New York: Fordham University Press, 2010), 20.
52. Ibid.
53. Heidegger, 'Insight into That Which Is', 27.
54. Ellul, *The Technological Society*, 125.
55. Anders, *Die Antiquiertheit des Menschen*, vol 2, 492.
56. Ibid.
57. Ibid., 130.
58. Arendt, 'Introduction *into* Politics', 109.
59. Ibid.
60. Jonathan Schell, 'In Search of a Miracle: Hannah Arendt and the Atomic Bomb', in *Politics in Dark Times*, ed. Seyla Benhabib (Cambridge: Cambridge University Press, 2010), 247–58.
61. Marcuse, 'World without a Logos', *Bulletin of the Atomic Scientists* 20 (January 1964), 25.
62. Dana Villa, *Public Freedom* (Princeton: Princeton University Press, 2008), 162.
63. Theodor Adorno and Max Horkheimer, *Dialectic of Enlightenment* [1947] (London: Verso, 1997), 121.
64. Arendt, 'Introduction *into* Politics', 108.
65. Jonas, *The Imperative of Responsibility*, 19; Mumford, *The Myth of the Machine: The Pentagon of Power*, 57.
66. Martin Heidegger, *Ponderings XII–XV* [1939–41], trans. Richard Rojcewicz (Bloomington and Indianapolis: Indiana University Press, 2017), 6.
67. Adorno and Horkheimer, *Dialectic of Enlightenment*, 9.
68. Arendt, 'On Violence', 112.
69. Ellul, *The Presence of the Kingdom*, 284–5.
70. Ibid., x.
71. Herbert Marcuse, *Eros and Civilization: A Philosophical Inquiry into Freud* (London: Ark, 1987), 5.
72. Ellul, *The Technological Society*, 277; Herbert Marcuse, *One-Dimensional Man* (Boston, MA: Beacon, 1964), x.

73. Marcuse, *One-Dimensional Man*, xvi.
74. Anders, *Die Antiquiertheit des Menschen*, vol 2, 492fn.
75. Ellul, *The Presence of the Kingdom*, 40; 48.
76. Heidegger, *Ponderings XII–XV*, 92.
77. Adorno and Horkheimer, *Dialectic of Enlightenment*, 104.
78. Heidegger, *Ponderings XII–XV*, 110.
79. Theodor Adorno, *Minima Moralia: Reflections from Damaged Life* [1951], trans. Edmund Jephcott (London: NLRB, 1974), 55.
80. Arendt, 'Introduction *into* Politics', 159–60.
81. Ellul, *The Presence of the Kingdom*, 80.
82. Hannah Arendt, 'Karl Jaspers' [1958], in *Men in Dark Times* (London: Cape, 1970), 82.
83. Ibid.
84. Ibid.
85. Ibid.
86. Ibid., 83.
87. Mumford, *The Myth of the Machine: The Pentagon of Power*, 256–7.
88. Ellul, *The Technological Society*, 116.
89. Ibid., 118.
90. Ellul, *The Presence of the Kingdom*, 17–18.
91. Mumford, *The Myth of the Machine: The Pentagon of Power*, 245.
92. Ibid.
93. Heinz Tschachler, '"Hitler Nevertheless Won the War": Lewis Mumford's Germany and American Idealism', *American Studies* 44:1 (1999), 97–8.
94. Mumford, *The Myth of the Machine: The Pentagon of Power*, 233.
95. Ellul, *The Presence of the Kingdom*, 287–8.
96. Ellul, *The Technological Society*, 289.
97. Mumford, *My Works and Days*, 5.
98. Stephen Bronner, *Reclaiming the Enlightenment: Toward a Politics of Radical Engagement* (New York: Columbia University Press, 2006), 95.
99. Ibid., 95–6.
100. Sheldon Wolin, *Fugitive Democracy* (Princeton: Princeton University Press, 2016), 218.
101. Arendt, 'On Violence', 147.
102. Arendt, 'Introduction *into* Politics', 110.
103. Ibid., 111.
104. Anders, *Die Antiquiertheit des Menschen*, vol. 2, 78–9.

4

TECHNOLOGIES OF PRODUCTION AND THE RISE OF THE MACHINE

INTRODUCTION

'There is a widely held belief', wrote Pierre Francastel in 1956, 'that the most momentous event of our times is the machine's sudden and absolute ascendency over the conditions of human existence.'[1] In due course, automation will eliminate manual work and greatly reduce the hours of work, argued philosopher and sociologist of industrialisation, Georges Friedmann, in the same year.[2] It was John Diebold's 1952 *Automation* that had first popularised the term. Automation forms a 'new chapter', he argued, in the 'story of man's organization and mechanization of the forces of nature'.[3] Diebold, an American businessman, was himself no critic of technology, but he recognised and wrote about the way automated technologies had transformed production, influenced the global economy and society, and were likely to continue to do so in the future. 'To a great extent,' he wrote, 'man's [productive] function is the tending of machines.'[4] Across factories and offices, economic activity was increasingly shaped by machines in fundamental ways, he explained, a trend that he believed would only continue with the coming of the computer age:

> The new technology makes possible the application of self-correcting, feed-back control on a widespread basis . . . This means that the mechanization of certain industries can now advance to a higher level. At the present level, human physical power is replaced by machine power.

At the higher level, the monitoring and control tasks now humanly performed will be done by machines.[5]

The changes Diebold identified in production processes were reflected, however – now as deeply troubling transformations – in the work of more critical observers. The concept of automation was pivotal to the critical understanding of technology outlined here, particularly with respect to the sphere of production or labour. The critics of technology characterise modern technologies of production as automated: they operate not as tools but as autonomous or semi-autonomous entities. Within the complex of 'automation' we can also include techniques of management and bureaucracy that support autonomous technology. Ellul is explicit that techniques of management form part of 'technique' as much as technologies themselves, and more broadly, the reaction of the critics of technology to automation in work should be set against the context of the proliferation of management theory and the development of scientific-industrial management techniques in the mid-twentieth century. The concept of 'autonomous technology' that has been widely recognised in this period (most notably outlined by Winner in *Autonomous Technology*) draws significantly from these trends in productive technology. Thus 'technology' as a concept is shaped in a fundamental way by productive technologies. In social and political terms, for the critics of technology, the entanglement of labour and production with modern technology entails the alienation, or deepening of the alienation, of the individual or labourer from the product, from the production process, and from the sense of purpose or meaning embodied in work. The individuals within such a technological society become, instead of producers, primarily consumers, and, as such, become alienated even from themselves through the manufacture of needs. For these thinkers, modern technological production must also be seen to exist in a close relationship with the technologies of destruction: the production of nuclear weapons is considered to be a transformative mode of 'making' for human productive capacity in various ways.

The ways in which technology has changed production *since* the Industrial Revolution provoked a distinct critique of Marx within the Frankfurt School. As Marcuse explained:

> Marx stressed the essentially 'neutral' character of technology . . . [that] modern machinery is susceptible to capitalist as well a socialist utilization. This amounts to saying that mature capitalism and socialism have the same *technical* base, and that the historical decision as to how this base is to be used is a *political* decision.[6]

As we have seen, Marcuse, along with Adorno and Horkheimer, rejected this claim of neutrality. 'The abolition of private property, does not, by itself, constitute an essential distinction as long as production is centralized and controlled over and above the population,' Marcuse writes.[7] Rather, the alienation of labour, and the fact that this *continues* (albeit reduced) even in the Marxian socialist society is, for Marcuse, more significant. The working day 'would remain a day of unfreedom, rational but not free'.[8] Anders also seeks to evolve Marxist ideas in the context of what he describes as the technological/totalitarian era. Arendt engages with Marxist ideas of labour from a more straightforwardly critical position in her description of the historical rise of labour in modern society, contrasting labour with the categories of work and action, to describe the mode of technological production in modernity. Marx and Marxism, for all the critics of technology, is the foundational intellectual protagonist for their thinking on production, yet all these theorists are in different ways profoundly critical of Marx's idea of technology's influence on politics and society.

One essential question here which helps to position the critics of technology in a broad sense against the background of Marxist ideology is what they understand the relationship between technology and capitalism to be in the modern world; whether technology is driven by capitalism, or whether the reverse is true. Is technology, or capitalism, the more significant force in the contemporary world, in other words? Delanty and Harris argue that Frankfurt School scholars, 'unlike the conservative critique of technology . . . saw a close connection between technology and capitalism', including as 'conservatives' Mumford and Ellul, as well as Heidegger, Gehlen and Spengler.[9] In fact *all* the critics of technology who feature in this work identify a close connection between modern technology and capitalism. None would simply subsume technological development under capitalism; as Delanty and Harris explain, the Frankfurt School thinkers, 'unlike much of later Marxist theory . . . did not see technology as simply reducible to capitalism but identified technology as having its own dynamics'.[10] Ellul goes so far as to argue that 'in our technological society, the forces of production are no longer the infrastructure. They have become a superstructure.'[11] For Arendt, capitalism and technology have risen side by side, although technology has now become the more significant force.[12] Heidegger writes that 'only modern technology makes possible the production of all these economic standing-reserves', a claim that suggests modern technology precedes the emergence of capitalism (the essence of technology, preceding, of course, machine technology itself).[13] Mumford distinguishes capitalism and technics, describing them as mutually influential tendencies. 'Although capitalism and technics must be clearly distinguished at every stage, one conditioned the other and reacted upon it.'[14] The 'megamachine', he argues, was the origin of 'civilization', oriented around 'mechanically organized work and mechanically organized destruction and extermination'.[15] For Marcuse, because control, not ownership of production, enables domination, capitalism alone (as a system of private property) does not explain production dynamics today. Capitalism may still arguably be more important as an explanatory factor in Marcuse's work, but technology enables the extension of control over society *even* in conditions of plenty, and thus enables a continuing, indeed worsening, condition of domination. From all perspectives, the Marxist analysis of capitalism gets technology, and the relationship of technology with capitalism, completely wrong.

Autonomous Technology in the Contemporary World

The concept of autonomous technology is drawn from these theorists' observations of the state of production technologies. Specifically, there is a distinct line of argument and set of observations in their writings on the manner in which technology has undergone a qualitative shift since the Industrial Revolution: a transformation from a technology of tools to a technology of autonomous machinery. Even those theorists who are less focused on labour technologies – such as Jonas – make this point. 'Modern technology, in the sense which makes it different from all previous technology, was touched off by the industrial revolution,' he writes. It 'determined the first distinctive feature of modern technology, namely the use of artificially generated and processed natural forces for the powering of work producing machines'.[16] For Mumford, Anders and Arendt, in particular, the shift from tool-technology to machine-technology features centrally in their analyses of modern technology. In this context, they ask, do the machines serve us, or do we serve the machines?

'The essential difference between a machine and a tool', Mumford explains, 'lies in the degree of independence in the operation from the skill and motive power of the operator: the tool lends itself to manipulation, the machine to automatic action.'[17] Modern economic systems utilising 'mechanization and automation . . . ignored the human premises upon which the older agricultural and handicraft technologies had been founded'.[18] Such a system grows 'more dehumanized in proportion to the completeness of its automation and its extrusion of the worker from the process of production'.[19] The outcome of 'universal automation' is 'boredom and suicidal desperation' in the displaced worker, Mumford argues, and 'abject dependence on the Megamachine'.[20]

'If someone uses an instrument, such as pliers, he does not serve the pliers,' writes Anders. 'On the contrary: he dominates them, because he uses them for the purpose of his work.' In the context of machine production lines, the relation is different. To the extent that workers play a role in the functioning of factory production (and even the most highly automated factories still require some workers, Anders notes), the factory owner 'uses them so that he can use his machines effectively, or rather, *he serves the workers to the machine for it to make use of them*'.[21] As such, Anders concludes, 'As pieces of this world of machines, we humans are in the best cases proletarians. But most probably we are something worse than that.'[22]

Arendt argues that the Industrial Revolution 'and the emancipation of labor replaced almost all hand tools with machines which in one way or another supplanted human labor power with the superior power of natural forces'. As an essentially 'conditioned' being, that is, one who is shaped by the conditions of their existence, 'man "adjusted" himself to an environment of machines the moment he designed them'.[23] This was true of tools, *yet* 'the case of machines is entirely different':

> Unlike the tools of workmanship, which at every given moment in the work process remain the servants of the hand, the machines demand that the laborer serve them, that he adjusts the natural rhythm of his body to their mechanical movement. This, certainly, does not imply that men as such adjust to or become the servants of their machines; but it does mean that, as long as the work at the machines lasts, the mechanical process has replaced the rhythm of the human body.[24]

The shift from tool to automated technology thus shapes the relationship of worker to technology; machine technology shapes the world of labour in ways prior technologies did not. It also evidently changes the nature of the labour that is undertaken: automation takes over roles and aspects of work that were formerly undertaken by labourers themselves, creating in turn new demands for labour to service the machines. The influence of automation on (human) labour is thus complex. Despite our theorists' scepticism about technology, they largely agree that there are at least superficially beneficial aspects to the rise of automated production, not least the less 'laborious' character of the labour humans undertake. Yet on a deeper examination, they argue, such technologies mask or create increasingly concerning trends in the sphere of production.

The question of the influence of automated technology on production is at the centre of Marcuse's *One-Dimensional Man*. Because, he argues, productive technologies free up human energy for tasks other than essential labour, it increasingly necessitates deeper and more insidious forms of social domination. Because automation is so efficient, it frees up the time and energy of citizens, including for potential challenges to existing social orders:

> Rationalization and mechanization of labor tend to reduce the quantum of instinctual energy channelled into toil (alienated labor), thus freeing energy for the attainment of objectives set by the free play of individual faculties. Technology operates against the repressive utilization of energy insofar as it minimizes the time necessary for the production of the necessities of life, thus saving time for the development of needs *beyond* the realm of necessity and of necessary waste.[25]

As such, he claims, in order to ensure the continued subservience of the people to the system, domination increases in proportion to technological advancement. 'The closer the real possibility of liberating the individual from the constraints once justified by scarcity and immaturity, the greater the need for maintaining and streamlining these constraints lest the established system of domination dissolve,' he writes. The productive resources of industrial society are therefore turned against its citizens, becoming an instrument of control. 'Totalitarianism spreads over late industrial civilization wherever the interests of domination prevail upon productivity, arresting and diverting its potentialities.'[26] The actual elimination of human labour from material production via automation would eventually, logically, bring about the end of capitalism, he writes.[27] Yet, the very tendencies that generated greater productive efficiency via mechanisation also provide the means for the continued and indeed accentuated domination of societies. 'Mechanization and rationalization generated attitudes of standardized conformity and precise submission to the machine ... rather than autonomy and spontaneity,' Marcuse argues.[28]

This agrees with and develops Adorno and Horkheimer's earlier observations that, while the 'apparatus provides for him as never before', the individual 'disappears before the apparatus which he serves'.[29] Twenty years later, Adorno wrote again on the increasing alienation of the labourer from the system of production:

> The false identity between the organization of the world and its inhabitants, an identity created by the expansion of technology, amounts to the affirmation of the relations of production, for whose beneficiaries we seek today almost as vainly as for the proletariats, who have become all but invisible. The system has now become independent, even of those who are in control.[30]

Arendt offers a very different but no less pessimistic narrative on the evolution of production and the labourer's role in production within a technological society. She distinguishes between 'work' – the goal-oriented production of durable goods that are 'worldly' because they make up the fabric of the physical world in which we live – and 'labour' – the necessary and cyclical activity of producing goods for basic, natural needs. Yet 'work is now performed in the mode of laboring', she complains.[31] This transformation of work into labour, or the eradication of the distinction between these activities, has been in progress for centuries. Technology acts to accelerate and also to mask this fact. Labourers are still essential to the production process, she observes. While 'tools and instruments ease pain and effort and thereby change the modes in which the urgent necessity inherent in labor once was manifest to all', she claims that 'they do not change the necessity itself; they only serve to hide it from our senses':[32]

> Although machines have forced us into an infinitely quicker rhythm of repetition than the cycle of natural processes prescribed – and this specifically modern acceleration is only too apt to make us disregard the repetitive character of all laboring – the repetition and the endlessness of the process itself put the unmistakable mark of laboring upon it.[33]

Indeed, Ellul argues the opposite is true: not only has technology not alleviated the laboriousness of work, 'work has actually become even more laborious, more draining than before'. Full automation is rare, he argues, and 'in reality, for most workers, technological growth brings harder and more exhausting work (speeds, for instance, demanded not by the capitalist but by technology and the service owed to the machine)'. Technology brings not leisure, he insists, but only a new and more complete form of alienation. 'Alienation . . . is no longer capitalistic, it is now technological.'[34]

The concept of labour has come to dominate modern society, Arendt writes, and labour has been transformed from an originally private activity – the necessary maintenance of one's self – into *the* public activity, and it is this merging of labour and the public, or the political, that Arendt terms 'the social sphere'. 'The private realm of the household was the sphere where the necessities of life, of individual survival as well as of continuity of the species, were taken care of and guaranteed,' she writes.

> One of the characteristics of privacy, prior to the discovery of the intimate, was that man existed in this sphere not as a truly human being but only as a specimen of the animal species man-kind. This, precisely, was the ultimate reason for the tremendous contempt held for it by antiquity. The emergence of society has changed the estimate of this whole sphere but has hardly transformed its nature.[35]

The modern world, with tendencies and trends accelerated and accentuated by technology, has seen a transformation of the public-political sphere into one characterised primarily by the mode of labour: prioritising the necessary, rather the free, the private and individual rather than the community, and where the mass rather than the citizen has come to dominate. 'The ideals of *homo faber*, the fabricator of the world, which are permanence, stability, and durability, have been sacrificed to abundance, the ideal of the *animal laborans*,' Arendt claims. 'We live in a laborers' society because only laboring, with its inherent fertility, is likely to bring about abundance.'[36] The alienation that Arendt identifies in modern production technology, and in the rise of labour that is interwoven with it, is the collapse of work and thus worldliness. Technological production maintains the essence of labour, but because of this it cannot be a replacement for work: technology cannot construct or maintain the worlds that enable political communities to function as communities.

Yet, even as she makes these claims, Arendt projects forward to a future in which the machine takes over all labouring tasks. 'The advent of automation . . . in a few decades probably will empty the factories and liberate mankind from its oldest and most natural burden, the burden of laboring and the bondage to necessity,' she writes. 'It seems as though scientific progress and technical developments had been only taken advantage of to achieve something about which all former ages dreamed but which none had been able to realize.'[37]

> However, this is so only in appearance. The modern age has carried with it a theoretical glorification of labor and has resulted in a factual transformation of the whole of society into a laboring society. The fulfilment of the wish, therefore, like the fulfilment of wishes in fairy tales, comes at a moment when it can only be self-defeating. It is a society of laborers which is about to be liberated from the fetters of labor, and this society does no longer know of those other higher and more meaningful activities for the sake of which this freedom would deserve to be won . . . What we are confronted with is the prospect of a society of laborers without labor, that is, without the only activity left to them. Surely, nothing could be worse.[38]

Both Anders and Jonas ask the same question: how does a society based upon labour fare once labour is replaced by machines? Technology today, argues Anders, increasingly attempts 'to make men superfluous: namely, for their work to be replaced by the automation of the machines; to realise a condition, in which . . . as few workers as possible are required'.[39] Yet the elimination of work would be truly 'hellish', he writes, because we would be 'cheated out of one of the most powerful, important and most-loved pleasures, namely the "*voluptas laborandi*" (mostly overlooked in view of the difficulty of work).'[40] Jonas points, too, to the emptiness of a post-work world. 'Work *per se* . . . [became] a "first need of life"', he writes, 'by having been taken away from man by the machine, whose work secures the fulfillment of his formerly "first" needs.' This produces the 'paradox that the new need for work – not for its yield – becomes perhaps the hardest to satisfy of all those needs whose satisfaction the new society is mandated to assure to each of its individuals'.[41]

Thus, as producers, or labourers, technology has transformed our experience of work, and according to these thinkers, significantly for the worse (even against the apparent easing of work's laboriousness, or the promise of such). But technology and mass production has also transformed the scope of production, or the scope of its domination over society. Production technology shapes not just work, but also, increasingly, consumption, and in modernity, we are characterised increasingly *as* consumers first and foremost.

For Arendt, it is the turn to labour from work that necessitates the consumption character of modern society. 'The endlessness of the laboring process is guaranteed by the ever-recurrent needs of consumption; the endlessness of production can be assured only if its products lose their use character and become more and more objects of consumption.'[42] Consumption relates to the prioritisation of the life-process itself, self-preservation, and the things that make self-preservation possible. Capitalism, write Adorno and Horkheimer, requires self-preservation to be the central concern. 'Whoever resigns himself to life without any rational reference to self-preservation would, according to the Enlightenment – and Protestantism – regress to prehistory.'[43]

But, in consumption, the rhythms and demands of technological production remain just as dominant. 'The more the process of self-preservation is effected by the bourgeois division of labor,' Adorno and Horkheimer write, 'the more it requires the self-alienation of the individuals who must model their body and soul according to the technical apparatus.'[44] Within 'the sphere of consumption', Adorno would later add, people have become mere consuming functions: 'appendages of the machinery'.[45] Arendt is in agreement with this sentiment, writing that, instead of 'utility and beauty . . . standards of the world', we now design and create products whose 'shape will be primarily determined by the operation of the machine'. While the basic function of such products are 'the human animal's life process . . . the product itself – not only its variations but even the "total change to a new product" – will depend entirely upon the capacity of the machine'.[46]

The real Industrial Revolution, Anders argued, was initiated at the moment machines began creating machines (the 'date is of no concern', he claims).[47]

> The mechanism of our industrial cosmos now consists in the manufacture of products (that is, means of production) which in turn produce products . . . etc., until one last machine ejects the 'final products' that are no longer means of production but means of consumption, that is, those that are to be consumed through their use, like bread or hand grenades.[48]

The role of humans in this process is limited to the beginning (the invention or design of the initial machine) and the end, as users or consumers. In this chain of technological production, 'it is not actually the products themselves that figure as the means of production, but our *acts of consumption*', he writes, and it is 'a truly shameful fact, that our role as humans has become limited to ensuring that through our consumption of products (for which we must also pay), the process of production continues'.[49]

It is in the nature of consumption that its demands are endless. The drive for self-preservation and man's 'needs' expand exponentially. This is fundamental to Arendt's characterisation of 'labour'; it is implicit in Adorno and Horkheimer's critique of bourgeois production. But the nature of 'needs' in the contemporary technological world, and their production by and for technology, is most extensively explored by Marcuse. In a world where technological production has the capacity to provide for all, thus freeing the energies of the individual, those individuals can be distracted by the production of new needs. Technological modernity creates new 'needs', which thus enable the continuation of domination. He describes the new 'technostructure of exploitation' as 'the growing productivity of labor constantly augmenting the wealth of commodities and services; the intensified meaningless work and performances required for producing, buying, and selling these goods and services; and the scientific control of consciousness and instincts, that is to say, domination through steered satisfaction and steered aggression'.[50] The excessive productive capacity of advanced technocapitalism has altered capitalism's mode of domination through the construction of new needs. People 'have innumerable choices, innumerable gadgets which are all of the same sort and keep them occupied and divert their attention from the real issue – which is the awareness that they could both work less and determine their own needs and satisfactions'.[51] By this method 'the technological veil covers the brute presence and the operation of the class interest in the merchandise'.[52] It is not technology itself, but specifically technology and capitalism, hand in hand, that enable the total and nearly invisible control over society, through 'servitude and toil' but also through the 'greater happiness and fun available to the majority of the population'.[53] Today, 'the majority of organized labor shares the stabilizing, counterrevolutionary needs of the middle classes, as evidenced by their behavior as consumers of the material and cultural merchandise', Marcuse argues.[54]

Thus leisure, too, is becoming, like work, shaped by automation. 'Amusement under late capitalism is the prolongation of work,' argue Adorno and Horkheimer. 'Mechanisation has such power over a man's leisure and happiness, and so profoundly determines the manufacture of amusement goods that his experiences are inevitably after-images of the work process itself.'[55] The 'television screen is the treadmill of leisure', wrote Anders. 'Our consumption must keep up with its tempo.' Leisure is today 'imposed on us just like work . . . a double unfreedom' which presents itself disguised as a double freedom.[56] Increasingly, agrees Ellul, 'the economic fact covers every human activity . . . [and] the development of techniques is responsible for the staggering phenomenon of the absorption by economics of all social activities'.[57] Arendt, using different phraseology, also outlines a similar transformation in leisure: the assimilation of leisure to labour. 'Culture relates to objects and is a phenomenon of the world; entertainment relates to people and is a phenomenon of life.'[58] In modernity, entertainment has expanded exponentially; by contrast, culture, as the title of the essay in which she writes these words asserts, is in crisis. 'An object is cultural to the extent that it can endure,' she explains; 'its durability is the very opposite of functionality, which is the quality which makes it disappear again from the phenomenal world by being used and used up.'[59] Against the worldliness of culture, entertainment is consumed, thus facilitating the endless cycle of labour through leisure.

The impact of technology on the production process has been transformative. Because of technology, the power and capacity of an economy's productive possibility have expanded, opening up both potential for freedom and the realisation of a more profound and extensive domination. The rhythm of production, or labour, has spread to every part of life, public and private: the political has become concerned primarily with economics, while the private sphere of leisure and entertainment has been infiltrated by the demands and influence of the production apparatus, where consumption is imposed upon the masses. Technology causes production to expand, but the changes it effects are qualitative in nature, not exclusively quantitative. Production technologies improve and worsen our condition as producers and consumers in myriad complicated ways, which are evolving all the time.

What can clearly be seen in these various claims about the influence of automated technology on production is the way in which production is increasingly alienated from the individual producer or labourer. Advanced industrial society, with 'its productivity and efficiency; its capacity to increase and spread comforts, to turn waste into need, and destruction into construction, the extent to which this civilization transforms the object world into an extension of man's mind and body makes the very notion of alienation questionable,' Marcuse notes. 'The people recognize themselves in their commodities; they find their soul in their automobile, hi-fi set, split-level home, kitchen equipment.'[60] Yet, in truth, this only constitutes 'a more progressive stage of alienation. The latter has become entirely objective, the subject which is alienated is swallowed up by its alienated existence. There is only one dimension, and it is everywhere and in all forms.'[61] The atomisation that characterises modern society is also *internal*, writes Adorno: 'No fulfilment may be attached to work, which would otherwise lose its functional modesty in the totality of purposes, no spark of reflection is allowed to fall into leisure time, since it might otherwise leap across to the workaday world and set it on fire.'[62]

AUTONOMOUS TECHNOLOGY AND THE CONCEPT OF REIFICATION

Georg Lukács's 1923 essay 'Reification and the Consciousness of the Proletariat' developed the idea of reification as he interpreted it in Marx's writings, a concept that would become influential in the work of contemporary and later Marxist theorists, particularly within critical theory, including

Adorno, Horkheimer, Marcuse and Anders. Their concept of technological production draws upon and develops Lukács' idea of reification, and even the work of theorists such as Heidegger and Arendt mirrors elements of the concept. Reification is a form of alienation, describing the commodity form in modern capitalism, where it has become 'the universal category of society as a whole', as well as the social effects of the domination of the commodity form.[63] The basis of the commodity structure is, writes Lukács, 'that a relation between people takes on the character of a thing and thus acquires a "phantom objectivity," an autonomy that seems so strictly rational and all-embracing as to conceal every trace of its fundamental nature: the relation between people'.[64] As relations become 'thing-like', man's own labour becomes 'something objective and independent of him, something that controls him by virtue of an autonomy alien to man'.[65] This has both an objective and a subjective aspect, explains Lukács:

> *Objectively* a world of objects and relations between things springs into being (the world of commodities and their movements on the market) . . . *Subjectively* – where the market economy has been fully developed – a man's activity becomes estranged from himself, it turns into a commodity which, subject to the non-human objectivity of the natural laws of society, must go its own way independently of man just like any consumer article.[66]

Within the process of reification as it advances in capitalist society, mechanisation plays an important role for Lukács, but always within a basically Marxist framework. He articulates Marx's distinction between labour and nature, explains Thomas Sheehan, taking the antithesis 'a step further' in describing 'the artificial world designed and constructed by human beings as a "second nature"'. This reified second nature includes, Sheehan writes, 'non-material artificial structures' within a technological space, such as labour exploitation.[67] From labour's transformation from craftwork to machine industry, we see 'a continuous trend towards greater rationalisation, the progressive elimination of the qualitative, human and individual attributes of the worker . . . With the modern 'psychological' analysis of the work-process (in Taylorism) this rational mechanisation extends right into the worker's "soul".[68] He contrasts contemporary society, based upon mechanical, rationalised production processes with earlier 'organic' production, which still 'bound individuals to a community'.[69] Mechanisation, by contrast, 'makes of them isolated abstract atoms whose work no longer brings them together directly and organically; it becomes mediated to an increasing extent exclusively by the abstract laws of the mechanism which imprisons them.'[70]

Marcuse, Adorno and Horkheimer, however, do not tend to view the reification that takes place through technology as a mere evolution of the capitalist dynamic but rather believe that the reification that takes place through contemporary technology is qualitatively distinct from that which preceded it, a position also echoed in the early work of Jürgen Habermas, and which had also been anticipated in Walter Benjamin's 1935 essay 'The Work of Art in the Age of Mechanical Reproduction'.[71] 'Benjamin's essays on technology and the Right were attempts to dissolve reification,' writes Herf, and 'Lukács's theory of reification as developed in history and class consciousness was a cornerstone of Benjamin's ideas on the aesthetics of technology in the German Right.'[72] It is not a rejection of Lukács (although the Frankfurt School, particularly Adorno, was critical of Lukács in various ways), more a shifting of emphasis onto the importance of technology in reification, and specifically automated technology. 'Only in the medium of technology,' writes Marcuse, 'man and nature become fungible objects of organization . . . Technology has become the great vehicle of *reification* – reification in its most mature and effective form.'[73] And automation, he argues, is *more* than simply the growth of mechanisation, 'it is a change in the character of the basic productive forces'.[74] Like Marcuse, Adorno believes 'that which determines subjects as means of production and not as living purposes, increases with the proportion of machines to variable capital'.[75] Everything, even thinking itself, is given up to the machine. 'Thinking objectifies itself to become an automatic, self-activating process; an impersonation of the machine that it produces itself so that ultimately the machine can replace it.'[76]

While Lukács writes of the autonomy of the thing or product, and the autonomy of labour from the labourer, or the alienation of man from both work and its products, the emphasis on autonomous technology in the Frankfurt School thinkers introduces a new dynamic into the concept of reification: labour and things are not only considered to be separate from the social relations they (genuinely) spring from, but actually become autonomous forces. Autonomous technology, like all contemporary labour and production, is, of course, still driven by the rationalising and dominating forces of capitalism, but nonetheless, these theorists consider this mode of reification to be distinct from that which came before (historically), and thus we can see that this is (intellectually) also distinct from Lukács's conceptualisation of reification.

It is worth noting another aspect of Adorno and Horkheimer's and Marcuse's thinking which develops 'reification' in a different direction. All three famously emphasise contemporary technological society's attempt to dominate nature absolutely. The attempt to unify knowledge for the purpose of mastering nature was the principle of the Enlightenment, argue Adorno and Horkheimer. 'As a technological universe, advanced industrial society is a *political* universe, the latest stage in the realization of a specific historical *project* – namely, the experience, transformation, and organization of nature as the mere stuff of domination,' writes Marcuse.[77] Mechanisation transforms the very nature of man, Adorno suggests. Mechanisation is not an influence on man from 'outside', he writes, not a mere deformation of man's character. Rather:

> there is no substratum beneath such 'deformations,' no ontic interior on which social mechanisms merely act externally: the deformation is not a sickness in men but in the society which begets its children with the 'hereditary taint' that biologism projects on to nature. Only when the process that begins with the metamorphosis of labour-power into a commodity has permeated men through and through and objectified each of their impulses as formally commensurable variations of the exchange relationship, is it possible for life to reproduce itself under the prevailing relations of production.[78]

The total transformation of men, and their total alienation from themselves, enables the continuous reproduction of the wholly reified individual: 'the coordination of people that are dead'.[79] Today, he writes, 'the truly bourgeois principle, that of competition, far from being overcome, has passed from the objectivity of the social process into the composition of its colliding and jostling atoms, and therewith as if into anthropology'; this is 'the subjugation of life to the process of production'.[80]

Günther Anders recognises the phenomenon of reification, too. In the present world, he writes, the worker 'no longer recognises the effect of his activity as his own, he no longer identifies with "his" product'.[81] However, he also claims that the concept of reification 'is no longer sufficient for labelling the situation today'. Instead, he suggests, 'we stand on the threshold of a new stage: a stage in which the form of the thing is avoided and the thing-form becomes fluid . . . For this fact, which has so far been neglected by theory, I propose the term "liquidation".'[82] When everything becomes objective, writes Anders, and one thereby 'neutralize[s] the difference between what is nearby and what is remote', there takes place an 'ascetic renunciation of the natural world perspective'.[83] In its place, the industries of radio and television 'are decisive for today's concept of world and object'.[84]

> Rather, radio and television produce a second world: the image of the world in which today's humanity believes it lives. And, in addition to this 'second' world, there is also a third world, namely the world of entertainment, in short: everything. And what is decisive is precisely that this 'everything' is no longer accepted as an object or property, but rather is fluid; or does not 'remain' at all, but in this fluid condition, in which it flows from the factory, it is 'introduced.' In fact, this occurs so smoothly that there can no longer be any question of an act of reception or even a conscious awareness of this.[85]

Anders claims that reification has also reached a distinct stage of its development in another way, as Christopher John Müller highlights. Contemporary man feels shame not because he has become reified, but the opposite. '*A new, second-level in the history of the reification of human beings* has been reached with this response of feeling *shame at not being a thing*. Humans now acknowledge the superiority of things . . . and welcome their own reification.'[86] For Anders, Müller concludes, 'machine-artifice configures our soul'.[87]

Reification, writes Hanna Pitkin, can be conceived 'in the relatively literal sense that human beings are the artificers, the res-making species. Hannah Arendt – though surely no Marxist – uses the word in precisely this way. Reification, she says, means "fabrication, the work of homo faber".'[88] There are affinities between Arendt's critique of the rise of labour and the shift to a consumer society, and the claims that Marcuse, Adorno and Horkheimer make about the technological quality of contemporary reification. The products of labour are by definition, for Arendt, commodities, consumables, not durable objects of work. Commodities are not 'world'-related products, which could encompass anything from buildings to artworks to poems, that is, objects that gather people together around or within an enduring physical space of reality. Commodities are individual, disposable and 'necessary', and modern autonomous technology extends this process of commoditisation. Commoditisation influences all things, social relationships, and also the character of humans themselves. The effects of commoditisation on the individual are consistently described in terms of automation. This is quite clearly the case for those who utilise the notion of reification, but even those theorists who do not, offer similar claims about the way man's nature or character is being transformed by modernity.

Georges Friedmann wrote that one of the most 'alarming' dangers of automated labour is '*the failure of human beings to participate* in an environment which they can now control from the outside by means of increasingly efficient, autonomous, and widespread techniques . . . Man . . . is acted upon passively . . . being influenced more and more by a "press-button" attitude.'[89] Likewise, Adorno claims mechanisation does not 'deform' from without but permeates individuals in the technological society. For Marcuse, 'mechanization and rationalization generated attitudes of standardized conformity and precise submission to the machine'.[90] Exploitation is 'transfigured' as the 'established values become the people's own values . . . the choice between social necessities appears as freedom'.[91] As we proceed, Anders argues, towards the totalisation of technologisation, where all persons are 'integrated . . . as functionaries', even today 'they are already seen as "candidates", as parts of the "universal machine" that is currently being developed'.[92] He concludes that we are swiftly moving towards a world where 'energy, things and men are all exclusively materials for potential requisition'.[93] The effect of this 'transformation of man into raw materials' can be observed, he believes, in Auschwitz, where the process was realised in its appalling reality.[94] Even Heidegger's analysis of the nature of technology mirrors this, in its transformation of nature – and, to a degree, also man – into 'standing-reserve'. For Heidegger, writes Zimmerman, 'in the technological age . . . what it means for something "to be" means for it to be raw material for the self-enhancing technological system'.[95] One of the 'essential moments in the way of being of contemporary being', argued Heidegger:

> is replaceability, the fact that – in a game that has become universal and where anything can take the place of anything else – every being becomes essentially *replaceable*. The industry of 'consumer' products and the predominance of the replacement make this empirically obvious. Today being is being-replaceable.[96]

Mumford describes the transformation of the individual in terms of the loss of the 'higher functions'. In *Technics and Civilization,* he outlines the emergence of 'Economic Man' in the Paleotechnic age (that is, the industrial era), describing them as 'penny-in-a-slot automaton . . . [a] creature of bare rationalism' who sacrificed 'most of the normal pleasures and delights of civilized existence to the untrammelled pursuit of power and money'.[97] While Mumford would later abandon some of the claims he made in this pre-war (1934) text, the idea that civilised man had sacrificed his autonomy and freedom remained throughout his work. At the moment man 'boasts of conquering nature, he surrenders his higher capacities, and he weakens his ability . . . for co-ordinated thought and disciplined action'. Thus, he concludes: 'today it is man's higher functions that have become automatic and constricted and his lower ones that have become spontaneous and irrational.'[98] Such a civilisation produces 'mass man: incapable of choice, incapable of spontaneous, self-directed activities: at best patient, docile, disciplined to monotonous work to an almost pathetic degree, but increasingly irresponsible as his choices become fewer and fewer'.[99] Both Mumford and Arendt identify that, in the subsumption of all activities to labour, only the lowest part of man, the animal component of his psyche, remains – in Arendt's terms, the *animal laborans.*

CAN LABOUR IN THE ERA OF AUTOMATION BE EMANCIPATED?

Autonomous productive technology is profoundly and essentially alienating, according to the critics of technology. It alienates people from one another and labourers from their product. The human world is dehumanised, as technology forms a second, false world – production is no longer intended to be for the benefit of consumers, nor to ease the burden of labour, but rather, increasingly labour and human nature are bent towards the end of production, specifically that of technological production. What distinguishes modern production technologies from earlier means of production is the autonomous quality of such technologies: the transformation from tool to machine. This has significant effects on labour and production: the machine's autonomy alienates the process of labour and the products of labour, and increasingly alienates humans from core elements of their own condition or psyche. While some of the theorists of technology are sceptical about the future potential for the genuine emancipation of labour, seeing only a likelihood of the deepening of existing tendencies, others are more hopeful. Adorno and Horkheimer are famously cynical about the possibility of emancipatory revolution, for example, while, for Arendt, a society based exclusively upon labour (rather than work) is profoundly atrophied, even more so if future technologies deny citizens even the possibility of labouring. Others disagree. Heidegger believes that, despite the fact that to date it has obscured the truth of being, technology may be the very path for the recovery of an understanding of being, *if* we can come to understand it, and our relationship to it, authentically. And Marcuse finds in technology the possibility for the development of labour for a more meaningful existence. All, however, imply that only radical change can transform contemporary technologised society – the potential of, for example, labour movements or more gradualist approaches are largely ignored or dismissed.

Marcuse, writes Feenberg, 'is a utopian thinker. He conceives of a redeemed technological rationality in a liberated society, much as Plato, at the end of the *Gorgias*, imagines a reformed rhetoric that would serve good ends.'[100]

> Eventually this utopianism enters into his conception of instrumental rationality itself where it is formulated as a positive technological alternative. Technology is to be reconstructed around a conception of the good, in Marcuse's terminology, around life. The new technical logos must include a grasp of essences, and technology must be oriented toward perfecting rather than dominating its objects. Marcuse thus demands the reversal of the process of neutralization in which technological rationality was split off from substantive rationality and subserved to domination.[101]

The automobile or the television, and the various gadgets of contemporary life, are not themselves repressive, Marcuse writes, but become so when they 'become part and parcel of the people's own existence, own "actualization"'.[102] While technological society tends to be totalitarian, he argues that this can be reversed. The machine's power 'is only the stored-up and projected power of man', and thus, he claims, 'to the extent to which the work world is conceived of as a machine and mechanized accordingly, it becomes the *potential* basis of a new freedom for man'.[103]

The error that Marx made, claims Marcuse, was to argue that labour – even in the socialised society – remains alienated labour, that 'the only thing that

TECHNOLOGIES OF PRODUCTION

can happen within it is for labor to be organized as rationally as possible and reduced as much as possible'. Labour thereby remains 'in and of the realm of necessity and thereby unfree'. By contrast, Marcuse writes, productive technologies open up new potentials, 'of letting the realm of freedom appear within the realm of necessity – in labor and not only beyond labor'.[104]

> the development of the productive forces beyond their capitalist organization suggests the possibility of freedom within the realm of necessity. The quantitative reduction of necessary labor could turn into quality (freedom), not in proportion to the reduction but rather to the transformation of the working day, a transformation in which the stupefying, enervating, pseudo-automatic jobs of capitalist progress would be abolished . . . This is the utopian concept of socialism which envisages the ingression of freedom into the realm of necessity.[105]

While technology increases alienation, Marcuse suggests, this development could in fact become genuinely progressive. 'The very progress of civilization under the performance principle has attained a level of productivity at which the social demands upon instinctual energy to be spent in alienated labor could be considerably reduced,' he writes.[106] 'Progressive alienation increases the potential of freedom: the more external to the individual the necessary labor becomes, the less does it involve him in the realm of necessity.'[107] Mumford, too, emphasises the possibilities of technology not in liberation from work, but the transformation of work to something 'more educative, mind-forming, self-rewarding work, on a voluntary basis . . . through a life-centered technology . . . an indispensable counterbalance to universal automation', in order to recover 'those parts of his personality that have been crippled or paralyzed by insufficient use'.[108] He appeals to 'those few individuals who, having resisted the industrial, bureaucratized, and specialized world, are capable of rejecting its conformist mechanisms and automatisms', writes Casillo.[109]

For Marcuse, this is largely an aesthetic revolution. For Marcuse, 'release from [technological] . . . experience can only come through another form of experience, an aesthetic form', writes Feenberg. In Marcuse's later works, Feenberg observes, he expresses that 'an authentic human existence is to be achieved at the level of society as a whole through the aestheticization of technology, that is, through its transformation into an instrument for realizing the highest possibilities of human beings and things'.[110] This reflects, in some ways, Benjamin's earlier claim that the technology of image reproduction offered, as Tyrus Miller describes, 'a potential to charge art with a new emancipatory function . . . the political "training" of the masses'.[111] On the other hand, while Marcuse recognises the potential for such an evolutionary development, he remains sceptical about its chances. 'Once the productive apparatus, under repressive direction, has grown into an apparatus of ubiquitous controls, democratic or authoritarian, the chances of a humanistic reconstruction are very poor.'[112] In the analysis of productive technology that the critics of technology offered, and especially the automated modern technologies that characterised the contemporary work sphere, a new kind of labour built a new kind of world, one in which humans were, increasingly, merely raw material.

WORK AND THE WORLD

Work, understood in a broad sense, has always been a central form of activity through which human beings interact with one another and form relationships, and through which they interact with the world of physical things and create that world. That is not to say that work, prior to the modern era, was a neutral or universally beneficial activity. Work activities and relationships are and have always been riven with sometimes egregious inequalities and injustices, and the way the world has been 'created' through work in different ways – not least the creation of wealth-driven structural injustices, but also the physical manipulation of the world – has been justifiably subject to endless criticism. The critics of technology believe that work, and thus its impact upon human relationships with each other and the world, is changing in novel ways. The increasing capacity of machines to automate production *could* produce a world in which expanded leisure time could be used for more worthwhile, humanistic activities, as Marcuse contemplates. Thinking in historical terms, the expansion of groups of people with an abundance of leisure time could conceivably mirror the capacity of groups in ancient (slave-owning) societies to dedicate that time to political activity, but now on a broader base of participation, and perhaps even without an underclass. For the most part, the critics of technology do not reach these conclusions, but rather more pessimistic ones about how the transformation and automation of work are in the process of changing the world. We can see this in at least three ways in the work of the critics of technology: the way work and workers are valued in the machine age; the way the relationship between work and the physical world changes in an era of automation; and the way that a transformed world of work changes the relationships between people as producers.

First, the absence of work or the possibility of the absence of work in a fully automated future is seen to undermine the value of human labour and thus humans themselves. Whether these theorists believe work or labour is simply part of the human condition (as is the case for most of these thinkers) or, as in the case of Arendt, believe that contemporary society (problematically) values work or labour above all other activities, the prospect of a workless world can only lead to the devaluation of humans *as workers*. This is particularly accentuated by contemporary consumer society where production is valued particularly in terms of consumption–production cycles. As Heidegger writes, being becomes being-replaceable. Where the machine undertakes all work, or automates and organises the vast majority of work, human workers have value only in relation to that order. The traditional Kantian relationship between means and ends has been upturned: humans here have value only *for* the machine, the labour and value producers. Work is not necessarily always harsher, writes Ellul, but modern work inevitably demands different qualities in the worker: 'It implies in him an absence, whereas previously it implied a presence. This absence is active, critical, efficient; it engages the whole man and supposes that he is subordinated to its necessity and created for its ends.'[113]

Second, the combination of consumerism and automation leads to the destabilisation of the 'world' as a physical artefact embodying historical, cultural and aesthetic facets; the things constituting the world become increasingly ephemeral, while the technological system becomes increasingly dominant. The physical world also becomes no longer 'ours' but rather made by and for the machine. In automation, humans lose their physical connection and ownership over the world. 'It has often been said that work is what makes the human being the mediator between nature and mankind as a species,' wrote Ellul. Today, technology is the *only* mediator, he argues, and both 'escapes any system of values' and also creates for itself 'the most immense set of mediations imaginable'.[114]

Finally, the automation of work, comprising both transformed attitudes towards human value, and the transformation of human relationships towards the physical world of things, entrenches and extends the massification of human societies. Technologised society is mass society: increasingly atomised and bereft of natural relationships, whether work (in this case) or, increasingly, all human relationships.

Notes

1. Pierre Francastel, *Art and Technology in the Nineteenth and Twentieth Centuries* [1956] (New York: Zone Books, 2000), 29–30.
2. Georges Friedmann, *The Anatomy of Work* [1956] (London: Heinemann Educational Books, 1961), xvii.
3. John Diebold, *Automation* [1952] (New York: Amacom, 1983), 158.
4. Ibid., 65.
5. Ibid., 140.
6. Herbert Marcuse, *Eros and Civilization: A Philosophical Inquiry into Freud* [1955] (London: Ark, 1987), 185.
7. Ibid., 81–2.
8. Herbert Marcuse, *An Essay on Liberation* (Boston, MA: Beacon Press, 1969), 20–1.
9. Gerard Delanty and Neal Harris, 'Critical Theory and the Question of Technology: The Frankfurt School Revisited', *Thesis Eleven* 166:1 (2021), 90.
10. Ibid.

CATASTROPHIC TECHNOLOGY IN COLD WAR POLITICAL THOUGHT

11. Jacques Ellul, *The Technological System*, trans. Joachim Neugroschel (New York: Continuum, 1980), 64.
12. Hannah Arendt, *The Human Condition* (Chicago: University of Chicago Press, 1998), 257.
13. Martin Heidegger, *Four Seminars* (Bloomington and Indianapolis: Indiana University Press, 2003), 62.
14. Lewis Mumford, *Technics and Civilization* (San Diego, New York and London: Harcourt Brace Jovanovich, 1963), 26.
15. Lewis Mumford, *The Myth of the Machine: Technics and Human Development* (New York: Harcourt Brace Jovanovich, 1967), 221.
16. Hans Jonas, 'Seventeenth Century and After: The Meaning of Scientific and Technological Revolution', in *Philosophical Essays* (New York and Dresden: Atropos Press, 1980), 76.
17. Mumford, *Technics and Civilization*, 10.
18. Lewis Mumford, *The Myth of the Machine: The Pentagon of Power* (New York: Harcourt Brace Jovanovich, 1970), 142.
19. Ibid., 143.
20. Lewis Mumford, 'Technics and the Nature of Man', *Technology and Culture* 7:3 (1966), 84.
21. Günther Anders, *Die Antiquiertheit des Menschen*, vol. 2: *Über die Zerstörung des Lebens im Zeitalter der dritten industriellen Revolution* (Munich: Verlag C. H. Beck, 2018), 79.
22. Ibid., 121.
23. Arendt, *The Human Condition*, 147.
24. Ibid.
25. Marcuse, *Eros and Civilization*, 93
26. Ibid.
27. Herbert Marcuse, *Counterrevolution and Revolt* (Boston, MA: Beacon Press, 1972), 3.
28. Marcuse, *Eros and Civilization*, 84.
29. Theodor W. Adorno and Max Horkheimer, *Dialectic of Enlightenment* [1944/47] (London: Verso, 1997), xiv–xv.
30. Theodor Adorno, 'Late Capitalism or Industrial Society?', in *Can One Live after Auschwitz?: A Philosophical Reader* (Stanford: Stanford University Press, 2003), 124–5.
31. Arendt, *The Human Condition*, 231–3.
32. Ibid., 125.
33. Ibid.
34. Ellul, *The Technological System*, 72–3.
35. Arendt, *The Human Condition*, 46.
36. Ibid., 124–6.
37. Ibid., 4.
38. Ibid., 4–5.
39. Anders, *Die Antiquiertheit des Menschen*, vol. 2, 28.
40. Ibid., 29.

41. Hans Jonas, *The Imperative of Responsibility* (Chicago: University of Chicago Press, 1984), 193.
42. Arendt, *The Human Condition*, 125.
43. Adorno and Horkheimer, *Dialectic of Enlightenment*, 30.
44. Ibid.
45. Theodor Adorno, 'Progress or Regression', in *History and Freedom: Lectures 1964–65* (Cambridge: Polity Press, 2006), 5–6.
46. Arendt, *The Human Condition*, 152.
47. Anders, *Die Antiquiertheit des Menschen*, vol. 2, 15.
48. Ibid., 15–16.
49. Ibid., 16.
50. Herbert Marcuse, 'Marxism and the New Humanity: An Unfinished Revolution', in *Marxism and Radical Religion: Essays Toward a Revolutionary Humanism*, ed. John C. Raines and Thomas Dean (Philadelphia: Temple University Press, 1970), 6.
51. Marcuse, *Eros and Civilization*, 100.
52. Marcuse, 'An Essay on Liberation', 14.
53. Ibid.
54. Ibid., 17.
55. Adorno and Horkheimer, *Dialectic of Enlightenment*, 137.
56. Anders, *Die Antiquiertheit des Menschen*, vol. 2, 115–16.
57. Ellul, *The Technological* Society, 158.
58. Hannah Arendt, 'The Crisis in Culture', in *The Promise of Politics* (New York: Schocken Books, 1993), 204.
59. Ibid.
60. Herbert Marcuse, *One-Dimensional Man* (Boston, MA: Beacon, 1964), 9.
61. Ibid., 11.
62. Theodor Adorno, *Minima Moralia: Reflections from Damaged Life* [1951], trans. Edmund Jephcott (London: NLRB, 1974), 130.
63. Georg Lukács, 'Reification and the Consciousness of the Proletariat', in *History and Class Consciousness* (Cambridge, MA: The MIT Press, 1971), 86.
64. Ibid., 83.
65. Ibid., 87.
66. Ibid.
67. Stephen Sheehan, 'The Nature of Technology: Changing Concepts of Technology in the Early Twentieth-Century', *Icon* 11 (2005), 1–2.
68. Lukács, 'Reification and the Consciousness of the Proletariat', 88.
69. Ibid., 90.
70. Ibid.
71. Robin Celikates and Rahel Jaeggi, 'Technology and Reification: "Technology and Science as 'Ideology'"', in *The Habermas Handbook*, ed. Hauke Brunkhorst, Regina Kreide and Cristina Lafont (New York: Columbia University Press, 2017), 260.
72. Jeffrey Herf, *Reactionary Modernism: Technology, Culture, and Politics in Weimar and the Third Reich* (Cambridge: Cambridge University Press, 1984), 31.
73. Marcuse, *One-Dimensional Man*, 168–9.

74. Ibid., 35.
75. Adorno, *Minima Moralia*, 229.
76. Adorno and Horkheimer, *Dialectic of Enlightenment*, 25.
77. Marcuse, *One-Dimensional Man*, xvi.
78. Adorno, *Minima Moralia*, 229.
79. Ibid.
80. Ibid., 27.
81. Anders, *Die Antiquiertheit des Menschen*, vol. 2, 73.
82. Ibid., 61–2.
83. Günther Anders, *Die Antiquiertheit des Menschen*, vol. 1: *Über die Seele im Zeitalter der zweiten industriellen Revolution* [1956] (Munich: Verlag C. H. Beck, 2010), 123.
84. Anders, *Die Antiquiertheit des Menschen*, vol. 2, 60.
85. Ibid.
86. Günther Anders, 'On Promethean Shame', in *Prometheanism: Technology, Digital Culture and Human Obsolescence*, ed. Christopher John Müller (London and New York: Rowman & Littlefield, 2016), 35.
87. Christopher John Müller, 'Introduction', in *Prometheanism: Technology, Digital Culture and Human Obsolescence* (London and New York: Rowman & Littlefield, 2016), 11.
88. Hanna Pitkin, 'Rethinking Reification', *Theory and Society* 16:2 (1987), 268–9.
89. Friedmann, *The Anatomy of Work*, 156.
90. Marcuse, *Eros and Civilization*, 84.
91. Marcuse, 'An Essay on Liberation', 36.
92. Anders, *Die Antiquiertheit des Menschen*, vol. 2, 123.
93. Ibid., 124.
94. Ibid., 23.
95. Michael E. Zimmerman, *Heidegger's Confrontation with Modernity: Technology, Politics, Art* (Bloomington and Indianapolis: Indiana University Press, 1990), xiv.
96. Heidegger, *Four Seminars*, 62.
97. Mumford, *Technics and Civilization*, 177.
98. Lewis Mumford, *The Conduct of Life* (New York: Harcourt, Brace & Co., 1951), 11.
99. Ibid., 16.
100. Andrew Feenberg, *Heidegger and Marcuse: The Catastrophe and Redemption of History* (New York: Routledge, 2005), 88.
101. Ibid., 88–9.
102. Marcuse, 'An Essay on Liberation', 14.
103. Marcuse, *One-Dimensional Man*, 2.
104. Herbert Marcuse, 'The End of Utopia', in *Five Lectures* (Boston, MA: Beacon, 1970), 63.
105. Marcuse, 'An Essay on Liberation', 20–1.
106. Marcuse, *Eros and Civilization*, 129.
107. Ibid., 222.
108. Mumford, 'Technics and the Nature of Man', 84.

109. Robert Casillo, 'Lewis Mumford and the Organicist Concept in Social Thought', *Journal of the History of Ideas*, 53:1 (1992), 113.
110. Feenberg, *Heidegger and Marcuse*, xiii.
111. Tyrus Miller, *Modernism and the Frankfurt School* (Edinburgh: Edinburgh University Press, 2014), 16.
112. Herbert Marcuse, 'Socialist Humanism', in *Socialist Humanism: An International Symposium*, ed. Erich Fromm (Garden City: Doubleday, 1965), 115.
113. Ellul, *The Technological Society*, 320.
114. Ibid., 36; 34–5.

5

THE VEIL OF TECHNOLOGY: MEDIA, PROPAGANDA AND IDEOLOGY

INTRODUCTION

In the early twentieth century, new technologies of radio, film and, later, television provided the platform for the emergence of truly mass media. These media brought the world into the home, operated continuously and without time lags, and exerted a degree of influence over individuals that was hitherto impossible. The new media transmitted unprecedented quantities of information as well as transforming the cultural landscape. By the period of the Second World War, the influence of media technologies had already effected widespread change in the way individuals related to the world around them, and the way that corporations, political parties, states and other groups interacted with individuals and society. Over the period of the Cold War, that influence, along with the prevalence of media technologies in society, grew. For some, this was a profoundly optimistic transition. Even Mumford wrote in 1934 (although his opinion later shifted) that 'instantaneous personal communication over long distances . . . is the mechanical symbol of those world-wide cooperations of thought and feeling which must emerge, finally, if our civilization is not to sink into ruin'.[1] In *Culture for the Millions*, a 1961 anthology of essays on mass culture in America, written by academics from a range of fields, 'the ringing defense of the mass media and their benefits by Lazarsfeld, Stanton, and Edward Shils . . . tends to predominate over and against those, like Arendt and Leo Löwenthal, who sound more cautionary notes'.[2] Alongside this, however, the critics of technology were expressing increasingly vehement opposition to the modern media industry.

142

THE VEIL OF TECHNOLOGY

A critique of the 'culture industry' was at the heart of Adorno's, Horkheimer's and Marcuse's work, as is well known. Given their concerns with the limitations of historical materialism as either an adequate means of understanding social change, or as an agent for radical change (as it appears in Marx's philosophy), the reason for their focus on the influence of media on individual consciousness (or the construction of false consciousness) is both clear and profoundly political in nature. Capitalist society is not bound to overcome itself through necessity, but rather, change depends upon the awareness and actions of the members of society. Contemporary media, under the control of the capitalist system, is an agent of capitalism, yet its influence is insidious, thanks in part to the mode by which contemporary media is delivered to its audience. 'These means', writes Adorno, 'especially radio and television . . . reach the people at large in such a way that they notice none of the innumerable technical intermediations: the voice of the announcer resounds in the home, as though he were present and knew each individual.'[3] What particularly distinguished the Frankfurt School's analysis from earlier Marxist theories, writes Martin Jay, was 'its refusal to reduce cultural phenomena to an ideological reflex of class interests'.[4] Rather, the influence of technology itself was centred: Adorno, for instance, criticised 'the deleterious effects of radio by pointing to its stimulus to standardization'. Although relating this to the permeation of the exchange ethic of capitalism, he also saw a connection with technological rationality itself.[5] The critique of mass culture, as Jay points out, had the greatest impact in American intellectual life of all the Institute's work.[6]

Less known in the anglophone world, but no less vociferous, is Anders's critique of the media industry, and the technologies that enable mass media. 'The events and objects of the world', he writes, are '"delivered to your home" through radio and television.' In this process, the outside world 'loses both its external character and its reality, so that we no longer encounter it as "world" . . . It is not a coincidence that totalitarian regimes of all shades have greedily seized upon the[se] instruments.'[7] Mass media was also a key topic for Ellul, not just in *The Technological Society*, but also in his 1961 text *Propaganda*, where he analyses in depth the impact of media and communication technologies on the actions of both democratic and totalitarian states in modernity. And Arendt also highlights the influence of media technology on contemporary society and politics, drawing connections between these technologies and their understanding of technology as such. For these thinkers, the medium most certainly is the message.

Certainly, their analyses of the impact of media technologies on modern society and politics reflect these theorists' broader understanding of the problematic nature of technology, but it is also not the case that their work on the influence of media is therefore superficial or purely derived from the overarching critique. While media technology, as a form of modern technology, is

143

seen to operate on the same broad principles that shape modern technology, media and communications technologies create their own specific issues and concerns in the political and social spheres. Broadly, media technologies are seen to repress freedom in contemporary societies in novel ways, progressing – across the globe – the totalitarian politics these thinkers are so fearful of. This chapter will focus on two ways in which they believe this happens: first, in the way that contemporary media technologies shape the world of human action, closing down spaces of freedom, including by shifting the relationship between the public and private; second, in the more conscious and overtly political use of propaganda by states and political authorities.

These philosophers are anxious about the way that technology is accelerating and enabling shifts in the way that public and private is understood, causing a collapse in the division between the two spheres. This has been most fully realised in totalitarianism but is also present in more moderate political regimes, as communication technologies enable the domination of public discourse and crowd out space for citizenship. This matters, because the norms that have governed the distinction between private and public activity have significant consequences for the potential for the realisation of freedom. Yet, despite their fears about technology, all concede that technology in itself does not definitively exclude the possibility of freedom. Technology could even, theoretically, contain the capacity to liberate. This is, however, *not* the case for technology in its modern form, which almost irresistibly and largely invisibly represses the possibilities for genuine political freedom.

Our theorists are also alarmed by the propagandistic quality of contemporary media technologies as a method of epistemic control, creating mass conformism and tendencies towards authoritarianism. The domination that results is not due to the political regime's guiding ideology (whatever that may be) but rather the more general influence of the technological behemoth and its essentially rational-efficient quality. Propaganda is 'totalitarian' but it is increasingly utilised by democratic as much as authoritarian leaders. This technology is political, but not partisan. This conceptualisation of a world equally shaped and thus corrupted by technology transcends Cold War divides of East and West. Liberal democracy cannot rescue the West, or the world, from technology, because like all contemporary political systems it is inextricably entangled with technology.

Yet, while these theorists critique the illusory 'freedoms' of contemporary democratic politics, they have rather different ideas of what freedom is. To some extent, it could be said that they are unified primarily by the identification of shared conditions of unfreedom, rather than a comprehensive positive conception of freedom. Yet, they do share a concept of freedom as dialectical, a claim that it embodies some form of tension which technology tends to repress. 'Ellul is a believer in genuine democracy,' writes Randal Marlin, 'but he thinks that the word "democracy" has been treated in a way that turns it into a myth rather than a reality.'[8] Ellul believes that liberty arises in opposing the constraints of contemporary political power, and ultimately in the religious transcendence of earthly constraints. Arendt identifies freedom with the uniquely spontaneous 'act', the ability to start anew, enacted into a political space of equality. Adorno and Horkheimer's understanding of freedom is more opaque. 'Freedom can only be grasped in determinate negation in accordance with the concrete form,' Adorno wrote; it can only be understood in relation to its social and historical context, but *because* men are always subject to the unfreedom of society, freedom can be found in resistance against society as it exists at that point in time.[9] The relationship between freedom and the individual is 'in large part identical' to the relationship of the universal to the particular, he argues.[10] Similarly, for Marcuse, freedom is understood dialectically. Freedom can only exist beyond the realm of necessity, but equally, the freedom to think relies upon a social and cultural world that makes it possible to construct ideas that challenge the existing social and political order.

In modernity, where the concept and reality of freedom have become impoverished, Marcuse's and Adorno and Horkheimer's analysis of freedom is expressed predominantly via their assessment of domination, and the closing down of the space for individual expression, resistance and thus dialectical social movement. Arendt's and Ellul's assessment of the status of modern freedom is equally cynical. Arendt argues that the very idea of freedom has largely been lost, and at best severely diminished, because the sphere of politics as a public, equal realm has been overwhelmed by the rise of sovereign political power.[11] Ellul believes that freedom is increasingly restricted by the technologised, bureaucratised systems that make 'us march like obedient automatons'.[12] The 'Automatism of Technical Choice' dictates that there is 'the one best way' and that 'technique itself, *ipso facto* and without indulgence or possible discussion, selects among the means to be employed'.[13] Anders is arguably the most pessimistic of all: 'the spirit of our time', he writes, 'is advancing towards total non-freedom . . . spontaneity is stifled in every domain, self-consciousness is extinguished, and there is an absolute absence of consciousness.' However, these critiques of modern political 'freedom' or its absence, despite their differences, rely upon a similarly fatalistic notion of technology.

Media and the Reshaping of the Political World:
Public and Private Spaces

The critique of media offers particular clarity as to these thinkers' concerns about the relationship between technology and freedom: revealing how the totalising and totalitarian qualities of technology in the abstract is realised through the development and utilisation of particular technologies in modern society. One of the ways in which these theorists agree that media technology is reshaping the political spaces of the modern world is its revision of the public and the private, and the boundaries between the two. This totalising movement is a major concern because it represents the consolidation of domination, and the shrinking of spaces for freedom. Concerns over the influence of the media on the public sphere and especially to democratic political culture are not new to these theorists, of course. Tocqueville, an important influence on Arendt, is one notable nineteenth-century analyst of the phenomenon, as were John Dewey and Walter Lippmann in the twentieth century. More specifically relevant to the question of technology's influence on society was the work of Walter Benjamin (perhaps, most importantly, 'The Work of Art in the Age of Mechanical Reproduction'), who was a figure close to Arendt and Anders as well as Adorno and Horkheimer. Nor did the critics of technology believe that the changing nature of the relationship between the public and private was exclusively a contemporary phenomenon; all identified marked changes as occurring throughout the modern era. Technology, therefore, does not create but entrenches a model of the political that leaves increasingly little room for freedom, they argue, and this can be seen in the development of media technologies and their increasingly significant influence upon the political sphere.

This is the case in totalitarian regimes, where intrusive and propagandistic technologies become a normalised part of political life. But it is also true, claim the critics of technology, in Western liberal-democratic regimes, where contemporary communications technologies allow political elites to monopolise public discourse (including *within* private spaces), radically shaping and delimiting public discourse, and effectively excluding citizens from this aspect of political participation. The development and use of communications technologies thus leads to the transformation of norms around the public/private distinction that has radical consequences for the possibility of freedom, a transformation that is all the more alarming for taking place almost invisibly.

Although Arendt did not base her ideas of public and private upon the influence of media technology, its relevance to her thinking is clearly discernible and at times made explicit in her writing. Increasingly, Arendt argues, the distinction between public and private has been degraded, merged into the 'social' sphere, where the public sphere is increasingly overwhelmed by a private sphere that has exceeded its traditional boundaries. 'In the modern world, the two realms . . . constantly flow into each other', to the extent that 'men have become entirely private . . . all imprisoned in the subjectivity of their own singular appearance'.[14] The public space that for Arendt provides a home for equality and freedom has largely disappeared in its original and authentically political form, as the private sphere, understood as the realm of the economic or life processes, comes to dominate politics. Relationships in this private sphere are characterised by *inequality*, the self-interested competition for scarce resources, and necessity takes precedence over free action.

Yet it is not simply that, as the private sphere expands, the public space shrinks. *Both* have evolved from their original forms. The public space has shrunk. But the private sphere is no longer simply the 'household' (the classical model Arendt refers to), or even its socialised modern equivalent, the economy, but has become 'a sphere of intimacy . . . whose peculiar manifoldness and variety were certainly unknown to any period prior to the modern age'.[15] More recently, totalitarianism has placed that version of the private realm under threat just as much as the public. *Both* public and private realms are under attack in the twentieth century, Arendt believed, because the conceptual boundary between the two has been erased by the rise of society, which today 'constitutes the public organization of the life process itself'.[16] That is, necessity governs society today, resulting in inequality and domination arising from the demands of the competitive 'life process'. The decline of the *private* sphere, no less than that of the public, is catastrophic. In Greece, Arendt writes, 'while to be political meant to attain the highest possibility of human existence, to have no private place of one's own (like a slave) meant to be no longer human'.[17]

Technology offers a kind of public unity in this world, 'for the first time in history all peoples on earth can have a common present,' she writes, 'no event of any importance can remain a marginal accident in the history of any other.'[18] Yet, we must not forget the other side of the coin, she argues: that 'the means of global communication were designed side by side with the means of possible global destruction'.[19] It is precisely because the public sphere has been eroded that technology can take on this mantle. Today's 'common factual present', unlike in earlier times, is not 'based on a common past' and it 'does not in the least guarantee a common future'.[20] For all its power, technology does not guarantee the durability of contemporary societies, rather, the concentration of extraordinary power in the hands of a very few individuals ensures the reverse. 'The solidarity of human life in this respect is entirely negative; it rests, not only on a common interest in an agreement which prohibits the use of atomic weapons, but, perhaps also . . . on a common desire for a world that is a little less unified.'[21]

Where Arendt tends to highlight the loss of the public sphere, Ellul's concern is with the erosion of the private sphere. Like Arendt, Ellul agrees that Western technology has negatively unified the world. 'Western rational, efficient methods have taken over the world . . . all peoples now confront one another directly,' he writes. 'The world has been shaped by mass media, a western invention.'[22] For Ellul, the key culprit behind the erosion of the private sphere is the modern state, which controls both public and private life, undermining what he considers to be a critically important dialectical tension between the two. 'Tension between groups composing the entire society is a condition for life itself,' he writes, 'the point of departure for all culture,' but also for the formation of the individual personality, and this tension must be sustained.[23]

The 'tension' of the Cold War international situation is inauthentic, he argues, because 'the aim is the exclusion or the elimination of one of the poles. In that case, development is unilateral.'[24] Only the reintroduction of a productive tension, he believes, possesses the capacity to restrict the now overwhelming power of the modern state. He critiques the 'illusion' that 'all problems have, in our time, become political', as the myth of contemporary politics.[25] Only the return of the public/private conflict, necessitating the return of the genuinely private spaces of life, can reinstate 'true reality'.[26] Private life needs to be 're-invented'.[27]

Anders also highlights the double-faceted shift that has taken place. 'Being at home in public', he writes, 'represents a feature of our contemporary existence that has so far been consistently neglected by theory.' While cultural critique has tended to emphasise the *deprivatization of the private sphere* . . . this only describes half of our current existence'. At the same time, and through the same processes, 'the *public sphere* has also lost its definition . . . [it] is often seen as *only a continuation of the private sphere*.'[28] This state of being has emerged through the technologies of production and particularly media, whereby the 'events and objects of the world' are 'delivered to your home'. The 'home' itself, formerly distinguished as a place of privacy, has been thoroughly invaded by the outside, such that the barrier between outside and inside, public and private, is indistinguishable, and the world accordingly loses its public quality.[29] When the masses are managed individually, conditioned *within* their homes without apparently losing their personality or rights, when the process is 'presented as *fun*', then such conditioning 'will succeed perfectly'.[30]

Adorno and Horkheimer do not attribute particular importance to the 'public' as a distinctly political realm. Yet, the changing dynamic between public and private does matter; they consider the *opposition* between them – an opposition which has largely disappeared – to possess enormous importance. Society inevitably exists, they write, but historically it was not inevitable that the individual would submit to society's demands. 'Once the opposition of the individual to society was its substance,' they write. 'It glorified "the bravery and freedom of the emotion before a powerful enemy, an exalted affliction, a dreadful problem"'. Today, in contrast, we have nothing but 'the nothingness of that false identity of society and individual'.[31] The individual as private person is erased in modernity because the culture industry and mass society have engulfed the space of privacy. And yet, like Arendt, Adorno and Horkheimer indicate that modernity is also characterised by a new type of privacy: 'Absolute solitude, the violent turning inward on the self, whose whole being consists in the mastery of material and in the monotonous rhythm of work, is the specter which outlines the existence of man in the modern world.'[32] This divides individuals from authentic community; their organisation is as commodities, not human beings.

148

While these authors characterise 'public' and 'private' differently, they agree that a certain dynamic that previously existed between two spheres has disappeared, to be replaced by an amorphous totality. They also agree that, while the (start of the) collapse of the public/private distinction preceded the technological era, the shape this relationship *now* takes is increasingly bound up with the development of modern technology.

Horkheimer writes that a reversal of public and private takes place through technologies such as 'photography, telegraphy and the radio'. Such technologies shrink the world, he claims, enabling us to 'witness the misery of the entire earth'. What was remote has become part of our world, and 'simultaneously, what is close has become the far-away. Now, the horror of one's own city is submerged in the general suffering, and people turn their attention to the marital problems of movie stars.'[33] He and Adorno describe how a monopolising, unitary culture turns all life into public life. 'Everybody must behave (as if spontaneously) in accordance with his previously determined and indexed level, and choose the category of mass product turned out for his type.'[34] Seemingly the most powerful technology in this transformation of culture which attacks the private is television, which 'plugs the gap' between 'private existence and the culture industry'.[35] Television, writes Adorno, 'obscures the real alienation between people and between people and things . . . They confuse what is mediated through and through – the life deceptively programmed for them – with the solidarity they are so acutely deprived of.'[36] While the despotism of culture might not be new – Villa highlights how Adorno and Horkheimer 'single out [Tocqueville's] critique of democratic culture as providing the leitmotif of their own' – the extent of its totalising power was new to the technological era.[37] 'Modern Communications Media have an isolating effect; this is not an intellectual paradox,' write Adorno and Horkheimer. 'Progress literally keeps men apart . . . Communication establishes uniformity among men by isolating them.'[38]

In his withering critique of contemporary technologised society in *One-Dimensional Man*, Marcuse argues that the conditioning of individual tastes, desires and opinions 'does not start with the mass production of radio and television and with the centralization of their control'. Rather, 'the people enter this stage as preconditioned receptacles of long standing'.[39] Nonetheless, he goes on to argue, 'private space has been invaded and whittled down by technological reality. Mass production and mass distribution claim the *entire* individual.'[40]

> The means of mass transportation and communication, the commodities of lodging, food, and clothing, the irresistible output of the entertainment and information industry carry with them prescribed attitudes and habits, certain intellectual and emotional reactions which bind the consumers more or less pleasantly to the producers and, through the latter,

> to the whole ... Thus emerges a pattern of *one-dimensional thought and behavior* in which ideas, aspirations, and objectives that, by their content, transcend the established universe of discourse and action are either repelled or reduced to terms of this universe.[41]

Marcuse echoes Adorno and Horkheimer's claims in the *Dialectic of Enlightenment*, says Dana Villa, when he writes that 'as a technological universe, advanced industrial society is a political universe, the latest stage in the realization of a specific historical project – namely, the experience, transformation, and organization of nature as the mere stuff of domination'.[42] Yet his image of contemporary media and culture is even more alarming and totalising than that offered in the *Dialectic*, Villa claims. 'Marcuse's conception of a "total society" – a society dominated by means of social control as ubiquitous as they are insidious (advertising; the culture industry; the "phony" freedom manifest in consumer and political choices) – consummates and codifies the dystopian potentials set forth in *Dialectic of Enlightenment*.'[43] Despite Marcuse's relative optimism with regards to the possible overcoming of contemporary technology's negative impact on society, he still asserts that the coordinated mobilisation of all media for its own defence has made it 'technically impossible' to utilise media for the purposes of speaking truth.[44] Media itself is irredeemably corrupt.

Ellul highlights the influence of radio and television on the erosion of the private sphere, both of which 'inserts the individual in the psychological and behavioral structure of the mass'.[45] The extension of technology into the individual's private world makes it increasingly political – 'what yet remains of private life must be forced into line by invisible techniques' – and it makes the political increasingly totalitarian, as it moves 'to absorb the citizen's lives completely'.[46] Anders also observes several similar consequences of the expansion of radio and television. They speak for us, he argues, depriving us of our ability to speak as individuals.[47] These technologies alter the private relationships of families within the home: unlike the 'centripetal force' of the table, which 'had encouraged those seated to share interests, glances and conversations, back and forth, to continue to weave the fabric of the family', the television screen is centrifugal. Family members, in front of the screen, no longer sit looking at one another, but side by side, facing the screen. Here, 'the possibility of seeing one another, of looking at one another, now only occurs by mistake, the ability to talk with one another (when one still can and will) only by accident'.[48] Finally, he argues that it is the 'purpose ... of the radio' to bring the outside world, or an image of it, into the house. '*Today we perceive "home" as the space that opens the door to the outer world to us* and brings us into real or supposed contact with it.'[49]

Anders also highlights that media communications operate in both directions: as we watch and listen to the outside world, so, too, do surveillance technologies enable states to watch and listen to what goes on inside the private spaces of the home. 'The devices are not only characteristic of totalitarianism; they are also "*totalising,*"' he insists.[50] This is important because it means that 'whether State A uses the devices because it is totalitarian, or whether State B becomes totalitarian because it uses the devices: it ultimately makes no difference.'[51] He points out that it is not just the fact of being watched, but the possibility of being surveilled at any time and in any place, that influences the behaviour of populations. The effect is already profound, he argues. 'Our lives have become universal property. Being asked about intimate details is considered completely acceptable, why we should not do or tolerate it is incomprehensible.'[52]

It also radically changes the nature of the spaces in which we coexist. 'The distance between the eavesdropper and the surveilled is now irrelevant,' he writes, and 'whereas earlier one was effectively not where one was not, and "being" had always meant "being in a certain place", one can now be in several places, and virtually everywhere at the same time'.[53] To exist at some distance from all others is the basis of individuation, writes Anders. 'The ontological fact of being unique means that each person, whether he wants it or not, represents an isolated *reserve*, shielded by walls, and as such, obstructs the claim to omnipresence and omnipotence of the totalitarian state.'[54] Where distance, individuation and uniqueness are progressively eliminated, totalitarianism thrives; it is able to become *truly* total.

Arendt argues that mass culture relies utterly upon modern technology, and that the culture of modern societies is now therefore largely determined by the nature of technology. The goods produced by the culture industry are mere 'consumer goods, destined to be used up', rather than cultural worldly objects of permanence.[55] This mass culture is neither public, because it is consumerist, ephemeral and driven by process, nor truly private, in the sense that it provides a space for intimacy or difference, because of its mass, standardised quality. Technology, she suggests, is destroying the very architecture of our world, breaking apart the remnants of the public sphere, but also the private as a space of difference, individuality and intimacy. It is only when the boundaries between worlds are protected and their differences recognised that freedom, the unique spontaneity of humans acting into the world, can be guaranteed.

Even Heidegger's highly abstracted understanding of technology in 'The Question Concerning Technology' and his earlier version of those claims specifically considers the influence of radio and film in modern society, albeit through the lens of his philosophical conceptualisation of technology. 'Radio and film belong to the standing reserve of this requisitioning through which the public sphere as such is positioned, challenged forth, and thereby first installed,' he writes. 'Their machineries are pieces of inventory in the standing reserve, which bring everything into the public sphere and thus order the public sphere for anyone and everyone without distinction.'[56]

These analyses of the relationship of public and private today are not identical. Importantly, these thinkers attribute different valuations to 'the public' and 'the private'. Arendt believes that the public is a necessary, highly beneficial space of politics, though largely lost in modernity. The concept is more critical in Ellul's, Marcuse's, Adorno's and Horkheimer's work: as the authoritarian 'state' for Ellul and a realisation of the capitalist system for Marcuse, Adorno and Horkheimer. Arendt connects the private sphere with the household, originally, and latterly with the economic sphere, identifying 'the private' as a space of intimacy as a relatively novel concept. In contrast, Ellul, Adorno and Horkheimer identify the private sphere primarily as a space of individuality and privacy.

Yet their ideas about the relationship between public and private intersect in significant respects. In particular, although their ideas of the public are quite different, these thinkers are closely aligned in their critique of the loss of the private sphere in modernity. They agree that modern politics and society do not permit that which was previously part of human life, a space in which people can exist meaningfully outside the public or political sphere, and all believe that tension between private and the public is important for holding open a space for freedom: for Arendt, largely because the public sphere cannot withstand the force of interests at play in the private sphere; for Anders, because our capacity to speak as private or public individuals has been lost; for Ellul, because only the tension between public and private can limit the modern state; for Adorno and Horkheimer, because only in the private sphere can individuals distinguish themselves from the systems that govern the world; for Marcuse, because the opposition of public and private comprises an essential part of recovering man's dialectical character. For all, a distinction between the two spheres of action is necessary for the realisation of freedom. Their collapse into one totality undermines freedom's potential.

This is in no small part due to modern technology, and there is a key parallel in these various portrayals of the contemporary 'private': while it is under attack on certain fronts, humanity is also increasingly imprisoned within a new sphere of the private. In modernity, technologies of media substantially exacerbate these trends. Technology plays a key role in repressing an older kind of private life, which is independently valuable, in favour of constructing a new kind of privacy or solitude that shapes society, and enables (totalitarian) politics to mould its subjects. This critique is especially visible in these theorists' critiques of the culture industry, and the technologies that enable and shape it.

PROPAGANDA AND TECHNOLOGY

For the critics of technology, the rise of propaganda is a key characteristic of modernity. Because propaganda is rooted in the conditions of modern existence, it can therefore be found, they suggest, in democratic as well as non-democratic societies. The technique of propaganda requires modern technology because it is 'mass' – creating and utilising mass society, and silencing agency. 'Propaganda never considers the individual as individual,' writes Ellul, 'but always in terms of what he has in common with others,' yet, acting upon the subconscious, it leaves 'the illusion of complete freedom'.[57] Yet, it 'strips the individual, robs him of part of himself and makes him live an alien and artificial life, to such an extent that he becomes another person and obeys impulses foreign to him'.[58] Propaganda is more than a particular tool, but 'the effect of a technological society that embraces the entire man and tends to be a completely integrated society'.[59] Arendt explored the way that totalitarianism specifically relies upon propaganda acting *upon* the mass, rather than the active adoption of totalitarian ideology.[60] Further, she argues that propaganda is a quality of modern communication itself. 'Not only political propaganda, but the whole of modern mass publicity contains an element of threat,' she writes.[61] Totalitarian propaganda 'perfects the techniques of mass propaganda', but did not invent them; these techniques emerged in the 'rise of imperialism and the disintegration of the nation-state'.[62] Horkheimer insists that the 'development of the mass media of propaganda such as newspapers, radio, television . . . must necessarily lead to dictatorship and the regression of humanity'.[63] The question of what compels obedience to propaganda is at the centre of Adorno's 1950 study of the authoritarian personality, as well as a major part of his earlier work on the Radio Research Project, which explored the effects of mass media on society. And as we have seen, Anders highlights the ways in which surveillance is equally responsible for the shaping of masses.

However, these theorists go further in arguing that not only are technology and propaganda intimately connected, but also the instruments of media technologies are *themselves* necessarily propagandistic: propaganda derives from the overarching logic of modern technology. Media technologies are essentially instruments of standardisation, Adorno and Horkheimer claim: the 'inherent tendency of the radio', for example, 'is to make the speaker's word, the false commandment, absolute', and thus 'turns all participants into listeners and authoritatively subjects them to broadcast programmes which are all exactly the same'.[64] This standardisation merges reality with illusion.[65] The only liberation such amusement offers 'is freedom from thought and from negation'.[66] Similarly, for Ellul, 'reality disappears . . . [because people] rely on printed paper, sound waves, or televised images. In their eyes, a fact becomes true when the newspaper prints an account of it.'[67]

The reason for its intrinsically propagandistic quality is that modern technology mediates the way individuals understand the world. Crucially, technology is behind the emergence and development of scientific reason, the critics of technology suggest, and *not* vice versa. Adorno and Horkheimer trace today's 'scientific attitude' to Bacon's work, asserting that 'technology is the essence of this knowledge', specifically, methods enabling exploitation.[68] What we aspire to take from nature is the ability to entirely dominate others, the pair argue.[69] Ellul, although he holds Enlightenment reason in higher regard than do Adorno and Horkheimer (or indeed Arendt), sketches its degradation into rationalism, 'incoherent, constricted, sectarian, and narrow-minded'.[70] Reason, 'invented' by the West, 'has been betrayed', he argues.[71] Rationalism is a form of knowledge that corresponds to technique: an increasingly illiberal doctrine that imposes itself upon the world, that seeks to control and dominate. Arendt writes that it was the essentially technological event of Galileo turning his telescope skyward that stimulated the rise of science.[72] Yet the rationality that emerged was 'a sort of brain power which in more than one respect resembles nothing so much as the labor power the human animal develops in its metabolism with nature', a form of thought, in other words, which reflected brute force. As labour extracts from nature what it needs, through violent means, so too does reason manipulate the material *it* works with: the ideas that make up our understanding of reality.[73] What is forced upon us, Anders submits, by the matrix of contemporary technology, is a 'totality . . . a pragmatic *image of the world* . . . to guide our actions, to shape our endurance, our behavior, our inaction, our taste, and therefore our entire praxis in general'.[74] Contemporary reason has the effect of constructing a self-realisation of the individual as part of a mass, a crucial enabler of propaganda. The proclaimed universality of reason claims dominance over the particular and differentiated thoughts or experiences of the individual, producing masses who act as undifferentiated crowds, and who can be managed as such. Modern masses, Arendt argues, are literally thoughtless; they do not believe their own senses, but their attention can be caught by anything 'at once universal and consistent in itself'.[75] The consistency of the system, not its relationship to reality, is what compels the mass and this enables propaganda's effectiveness. It also indicates the underlying power of technology itself. 'If it should turn out to be true,' writes Arendt, 'that knowledge (in the modern sense of know-how) and thought have parted company for good, then we would indeed become . . . thoughtless creatures at the mercy of every gadget which is technically possible, no matter how murderous it is.'[76]

Technique today dominates in the 'realm of intellect' just as in every other, writes Ellul.[77] Increasingly, propaganda and mass media communications make greater simplification and summarisation necessary, resulting in the level of the '"average reader" . . . automatically getting lower and lower'.[78] The incapacity for complexity in modern media – due to the fact that it is addressing crowds – operates according to a mechanised logic wholly in opposition to 'reason' in any traditional sense.

The collapse of thinking in mass society, and its relationship with the rise of modern technology, is also evident when Adorno and Horkheimer explain how, 'in the enigmatic readiness of the technologically educated masses to fall under the sway of any despotism . . . the weakness of the modern theoretical faculty is apparent'.[79] The effect of technology is, Horkheimer argues, to expand *mankind's* horizons, while at the same time diminishing the autonomy of the individual and their capacity to judge independently and resist propaganda and mass manipulation.[80] The political effect of this, as Brantlinger points out, is that, whenever Adorno and Horkheimer link the mass media to politics, 'the tie is always to fascism'.[81] The erosion of man's nature as an independently thinking creature is attacked by mass media, by its technologies, and the inevitable result is the rise of political fascism. 'Thinking objectifies itself to become an automatic, self-activating process; an impersonation of the machine,' they write, adding that this regressive trend 'finds its typical expression in cinema and radio'.[82]

This entrenched, essential relationship between technology and mass action is conveyed by Ellul in his claim that the purchaser of a television set 'obeys the collective motivations when he buys it, and through his act opens the door to propaganda'.[83] The audience for propaganda – 'the mass' – has already been prepared by technological processes. As John Omachonu and Kevin Healey write, for Ellul, mass media are a necessary precondition for propaganda. 'Without the media there can be no propaganda as it currently exists and, to become instruments of propaganda, the media must be centrally controlled on the one hand while showing on the other a "diversified" system of production.'[84] The control systems of the megamachine, says Mumford, 'are based, essentially, on one-way communication, centrally organized.'[85] The listener cannot answer back.

Media technologies in their contemporary form – a form which is essentially centralised 'broadcast' media – are thus essentially propagandistic, these thinkers argue. And while these theorists offer fuller analyses of this process than it has been possible to present here – exploring ideas of the state and sovereignty, of economic systems and processes, the influence of secularisation, among other important elements – it is not possible to explain the quality and influence of modern propaganda without reference to processes that they argue characterise contemporary 'technology', and their histories of the development of those technologies.

TECHNOLOGY AND FREEDOM

Contemporary technologised media shapes politics in fundamental ways, these authors argue. It does so by increasingly shifting and determining the nature of the political spaces that are available for us to occupy: traditionally private spaces are increasingly invaded; there is no distinction between the outside world and the inner sphere of the private life. Everything becomes, therefore, increasingly politicised, or subject to the political. Yet, at the same time, the traditional space of the public or political is itself shifting, according to these authors' overlapping claims: becoming 'social' rather than political, subject exclusively to economic demands, and increasingly a place only suitable for the 'mass' to occupy, as subjects, not citizens – reflecting a wider post-war 'artistic and philosophical preoccupation with authentic experience and the "end of the individual"'.[86] The normalisation of these changes, in large part through the means of new media, means the loss of the capacity to think any differently. This seemingly automatic and unconscious shaping of the fabric of the political world is accompanied by the much more conscious grasping of the power of media technologies by states and authorities, particularly in the form of propaganda, but also surveillance. These tools are so extraordinarily powerful that political elites are simply not able to resist the temptation to utilise them.

Standardisation and centralisation are both fundamental qualities of modern media technology, our theorists suggest. Ellul offers the particular example of the concentration of radio stations in France that occurred despite the grant of freedom to transmit anything. 'What outbreaks of joy there were when freedom of radio transmission was granted in France!' he wrote. 'It did not matter what ideas would be broadcast. Groups that had not been able to speak could now be heard.'[87] Yet audiences and funding were sparing, and ultimately these channels were forced off air. The same occurred in television programming, Ellul argues, 'a lowering of the cultural level, a debasement of the public, and an increasing advertising invasion'. Thus, he concludes, 'freedom brings about concentration'.[88] Certainly, the emphasis, or assumption, that is shared by the critics of technology was that mass media, and contemporary media technologies in general terms, were centralising, that is, increasingly operated from the centre by economic and political powers, and particularly, it is intimated, by the state. It was inherently propagandistic, and although other forms of propaganda *may* be found (Ellul famously distinguishes between vertical and the more unusual form of horizontal propaganda), 'vertical propaganda needs the huge apparatus of the mass media of communication'.[89] The technology of the culture industry, explain Adorno and Horkheimer, is 'no more than the achievement of standardisation and mass production . . . [and] the means which might resist central control has already been suppressed by the control of the individual consciousness'.[90]

For these theorists, technology does not wholly 'determine' politics. There *may* still be a way out. But because technology is so deeply enmeshed with politics in the twentieth century, it largely directs political action and its possibilities, severely restricting freedom. Our age presents a politics which is uniquely *unfree*, these thinkers contend, because the cumulative effect of technology is to erode our capacity to distinguish between freedom and domination. This occurs because technology has a fundamentally standardising, totalising effect – exemplified by media – which represses difference, distinction or dialectic, that is, the bases of freedom for these thinkers. Hence we find (again) that media technology, like modern technology in general, is not neutral but rather essentially totalitarian in this narrative: it creates a (false) world, over which it demands total dominance and thus the exclusion of free public, political, and, ultimately, also individual and private action. 'When technics becomes the universal form of material production,' Marcuse writes, 'it circumscribes an entire culture; it projects a historical totality.'[91] Domination *as* technology 'provides the great legitimation of the expanding political power, which absorbs all spheres of culture'.[92] All we can do, writes Ellul, is to set 'the Gorgon face of hi-tech' at a 'critical distance, for it is by being able to criticize that we show our freedom'.[93]

These theorists have different opinions on democracy as a form of government, ranging from the quasi-idealised (Arendt on classical democracy, Mumford on 'democratic' technics) to the rejection of the democratic tradition in the work of the Marxist-influenced theorists: Marcuse, Adorno, Horkheimer, Anders. Yet all agree that democracy is increasingly farcical. The very same technologically derived processes are occurring in both democratic and non-democratic nations: the production of the mass and mass opinion via advertising and media, surveillance and propaganda. What has not occurred in the West, it is suggested, is yet to come.

Differences between Western and Soviet society, Marcuse wrote in 1958, are paralleled by:

> [a] strong trend toward assimilation . . . centralization and regimentation supersede individual enterprise and autonomy; competition is organised and 'rationalized'; there is joint rule of economic and political bureaucracies; the people are coordinated by the 'mass media' of communication, entertainment industry, education.[94]

Under the 'the democratic or totalitarian aspects there is the reality of the technical state, which pursues its course regardless of whatever exterior form it may take', Ellul states.[95] And while, Anders argues, totalitarian governments have eagerly grasped at technological methods of control, *every* society that utilises these contemporary technologies, 'ends up in danger of also sliding into a political totalitarianism', as this technology leads toward the normalisation of the submission of the individual to its demands.[96] 'Nothing is more misleading than the . . . "philosophy of technology" which claims that devices are initially "*morally neutral*" and are thus freely available for any use: moral or immoral, humane or inhumane, democratic or antidemocratic.'[97] It is utterly impossible for people in this context to make reasoned judgements, possess rational views, or obtain the necessary knowledge to make free decisions, they argue, thus meaningfully democratic decision-making cannot take place. 'Democracy is based on the concept that man is rational and capable of seeing clearly what is in his own interest,' says Ellul, 'but the study of public opinion suggests this is a highly doubtful proposition. And the bearer of public opinion is generally a mass man, psychologically speaking, which makes him quite unsuited to properly exercise his right of citizenship.'[98]

However, the degree of equivalence between democratic and non-democratic states is disputed by these thinkers. In Anders's work, it often seems that there is very little to distinguish between the democratic West and other, more authoritarian states including those of Soviet Russia and even Nazi Germany. All are equally subject to the domination of the technological state. Marcuse, Adorno and Horkheimer agree that the Soviet state is no more free or democratic than the capitalist West, and indeed are profoundly critical of Soviet (and, of course, also fascist) governments. Ellul indicates that democracy is impossible in mass society, yet expresses a belief in democracy itself (albeit not in his current social context). And Arendt, while identifying anti-democratic trends in the West, believed, and argued for, the possibility of democratic resurgence in America, with its extraordinary democratic tradition.

The critique of mass media is a critique of the realisation of 'democracy' in the modern world. It is also a critique of a liberal idea of freedom. Negative freedom is simply not enough within a technological society where the people have become a mass, where thought and action has become standardised and its control increasingly centralised. The private sphere *cannot* be a space for freedom, and even more emphatically so in this context. Only within a society or political community where opposition, dissent, difference and individuality are given space to grow can such a freedom begin to be meaningful. This does not describe the context of their own world, these thinkers agree. For each of these thinkers, technology perverts not only modernity, but also what they believe might enable humans to escape their current condition: a positive tension or dialectic between different spaces or modes of activity: for example, for Ellul, technology is a modern (ir)religion, corrupting true spirituality by enforcing its singular ideology upon man; for Arendt, technology attacks the fragile 'public' intercourse of free human beings, destroying the possibility of creating a political space for action; for Adorno and Horkheimer, technology seeks to repress critique itself and thus the possibility of progress.

In his analysis of Marcuse's and Adorno and Horkheimer's critique of technological rationality, Dana Villa questions whether it is reasonable to 'posit – a priori, as it were – that this [political–public] sphere can never be the home of a critical or dissolvent rationality, a rationality that helps cut through illusions rather than simply underwriting their proliferation'.[99] Although he describes this approach as 'dubious', he also argues that they correctly identify at least certain aspects of the contemporary public sphere. 'It is, essentially, a "mediatized" public sphere, one in which advertising and the manipulation of symbols and popular resentments plays a far greater role than anything approximating rational – or even rhetorical – debate.'[100] Marcuse's concept of 'pre-conditioning', Villa continues, 'should be understood less as a form of brainwashing (an interpretation fostered by the bogus parallel between so-called "hard" and "soft" totalitarianism) than as what it unquestionably and irrefutably is: the shaping and manipulation of our desires by media'.[101]

The collapse of freedom is not a necessary outcome of all conceivable technological development. But the mode of technology that is distinctly modern stands in opposition to freedom because technology is itself so all-consuming and because it shapes an all-consuming politics that both isolates and radically 'socializes' (as a mass) in its pursuit of domination. Thinking in terms of technology is essential to understanding this collective claim that the character of modernity is essentially hostile to freedom, not least because the intellectual traditions these thinkers draw on do not necessarily explain their deep scepticism about freedom in the modern world: Arendt, the advocate of participatory politics, derides modern democracy; the Marxists Adorno and Horkheimer saw only a dead end where other socialists had seen an escape from capitalism through revolution; Ellul, the devout Christian, did not attempt to explain the collapse of freedom solely or even primarily in terms of processes of secularisation. Rather, their analyses of politics and the possibility of political freedom are rooted in an understanding of modernity in which totalitarian politics and technological processes together become predominant and mutually reinforcing processes.

NOTES

1. Lewis Mumford, *Technics and Civilization* (San Diego, New York and London: Harcourt Brace Jovanovich, 1963), 241.
2. David Jenemann, *Adorno in America* (Minneapolis: University of Minnesota Press, 2007), 37–8.
3. Theodor Adorno, *The Jargon of Authenticity* (London: Routledge and Kegan Paul, 1973), 76–7.
4. Martin Jay, *The Dialectical Imagination: A History of the Frankfurt School and the Institute of Social Research, 1923–1950* (Berkeley: The University of California Press, 1973), 178.
5. Ibid., 193.
6. Ibid., 217.
7. Günther Anders, *Die Antiquiertheit des Menschen*, vol. 2: *Über die Zerstörung des Lebens im Zeitalter der dritten industriellen Revolution* (Munich: Verlag C. H. Beck, 2018), 232; 239.
8. Randal Marlin, 'Jacques Ellul and the Nature of Propaganda in the Media', in *The Handbook of Media and Mass Communication Theory* (Chichester: John Wiley & Sons, 2014), 203.

9. Theodor Adorno, *Negative Dialectics* (London: Routledge, 1973), 231; Rolf Wiggershaus, *The Frankfurt School* (Cambridge, MA: MIT Press, 1994), 606.
10. Theodor Adorno, 'Progress or Regression', in *History and Freedom: Lectures 1964–65* (Cambridge: Polity Press, 2006), 5.
11. Hannah Arendt, 'What is Freedom?', in *Between Past and Future* (London: Penguin, 2006).
12. Jacques Ellul, *The Technological Bluff*, trans. Geoffrey Bromiley (Grand Rapids: Eerdmans, 1990), 411–12.
13. Jacques Ellul, *The Technological Society*, trans. by John Wilkinson (New York: Random House, 1964), 79.
14. Hannah Arendt, *The Human Condition* (Chicago: University of Chicago Press, 1998), 33; 58.
15. Ibid., 38.
16. Ibid., 46.
17. Ibid., 64.
18. Hannah Arendt, 'Karl Jaspers' [1958], in *Men in Dark Times* (London: Cape, 1970), 83.
19. Ibid.
20. Ibid.
21. Ibid.
22. Jacques Ellul, *The Betrayal of the West* (New York: Seabury, 1978), 123.
23. Jacques Ellul, *The Political Illusion*, trans. Konrad Kellen (New York: Knopf, 1967), 217; Jacques Ellul, *Propaganda: The Formation of Men's Attitudes*, trans. Konrad Kellen and Jean Lerner (New York: Knopf, 1965), 210–11.
24. Ellul, *Propaganda*, 219.
25. Ibid., 9.
26. Ibid., 12; 9.
27. Ibid., 206.
28. Anders, *Die Antiquiertheit des Menschen*, vol. 2, 95.
29. Ibid., 232.
30. Günther Anders, *Die Antiquiertheit des Menschen*, vol. 1: *Über die Seele im Zeitalter der zweiten industriellen Revolution* [1956] (Munich: Verlag C. H. Beck, 2010), 104.
31. Theodor Adorno and Max Horkheimer, *Dialectic of Enlightenment* [1944/47] (London: Verso, 1997), 153–4.
32. Ibid., 226.
33. Max Horkheimer, *Dawn and Decline* (New York: The Seabury Press, 1978), 19.
34. Adorno and Horkheimer, *Dialectic of Enlightenment*, 123.
35. Adorno, *Negative Dialectics*, 49–50.
36. Theodor Adorno, 'Prologue to Television', in *Critical Models: Interventions and Catchwords*, trans. H. W. Pickford (New York: Columbia University Press, 1998), 53.
37. Dana Villa, *Public Freedom* (Princeton: Princeton University Press, 2008), 156.
38. Adorno and Horkheimer, *Dialectic of Enlightenment*, 221–2.
39. Herbert Marcuse, *One-Dimensional Man* (Boston, MA: Beacon, 1964), 8.

40. Ibid., 10.
41. Ibid., 12.
42. Villa, *Public Freedom*, 161.
43. Ibid.
44. Marcuse, *One-Dimensional Man*, 68.
45. Ellul, *Propaganda*, 104.
46. Jacques Ellul, *The Technological Society*, 410–11; Ellul, *Propaganda*, 105.
47. Anders, *Die Antiquiertheit des Menschen*, vol. 1, 107.
48. Ibid., 106.
49. Anders, *Die Antiquiertheit des Menschen*, vol. 2, 92.
50. Ibid., 244.
51. Ibid., 244.
52. Ibid., 259.
53. Ibid., 247.
54. Ibid., 242.
55. Hannah Arendt, 'The Crisis in Culture', in *The Promise of Politics* (New York: Schocken Books, 1993), 202.
56. Martin Heidegger, 'Insight into That Which Is' [1949], in *The Bremen and Freiburg Lectures*, trans. Andrew J. Mitchell (Bloomington and Indianapolis: Indiana University Press, 2012), 36
57. Ellul, *Propaganda*, 7; Ellul, *The Technological Society*, 372.
58. Ellul, *Propaganda*, 169.
59. Ibid., xvii.
60. Hannah Arendt, *The Origins of Totalitarianism* (New York: Harcourt Brace Jovanovich, 1973), 341.
61. Ibid., 341fn.
62. Ibid., 351.
63. Max Horkheimer, *Dawn and Decline* (New York: The Seabury Press, 1978), 153.
64. Adorno and Horkheimer, *Dialectic of Enlightenment*, 159; 121–2.
65. Ibid., 126.
66. Ibid., 141.
67. Jacques Ellul, *The Presence of the Kingdom*, trans. Olive Wyon (Philadelphia: Westminster Press, 1951), 65–6.
68. Adorno and Horkheimer, *Dialectic of Enlightenment*, 4.
69. Ibid.
70. Ellul, *The Betrayal of the West*, 149.
71. Ibid., 148.
72. Arendt, *The Human Condition*, 261.
73. Ibid., 171.
74. Anders, *Die Antiquiertheit des Menschen*, vol. 1, 164.
75. Arendt, *The Human Condition*, 351.
76. Ibid., 3.
77. Ellul, *The Presence of the Kingdom*, 72.
78. Ibid., 67.
79. Adorno and Horkheimer, *Dialectic of Enlightenment*, xiii.

80. Max Horkheimer, *Eclipse of Reason* (London: Continuum, 1947), 5.
81. Patrick Brantlinger, *Bread and Circuses: Theories of Mass Culture as Social Decay* (Ithaca: Cornell University Press, 1983), 242–3.
82. Adorno and Horkheimer, *Dialectic of Enlightenment*, 25; xvi.
83. Ellul, *Propaganda*, 104–5.
84. John O. Omachonu and Kevin Healey, 'Media Concentration and Minority Ownership: The Intersection of Ellul and Habermas', *Journal of Mass Media Ethics* 24:2/3 (2009), 94.
85. Lewis Mumford, *The Myth of the Machine: The Pentagon of Power* (New York: Harcourt Brace Jovanovich, 1970), 183.
86. Stephen Bronner, *Reclaiming the Enlightenment: Toward a Politics of Radical Engagement* (New York: Columbia University Press, 2006), 115.
87. Ellul, *The Technological* Bluff, 110.
88. Ibid.
89. Ellul, Propaganda, 81.
90. Adorno and Horkheimer, *Dialectic of Enlightenment*, 121–2.
91. Marcuse, *One-Dimensional Man*, 154.
92. Ibid., 158.
93. Ellul, *The Technological Bluff*, 411–12.
94. Herbert Marcuse, *Soviet Marxism: A Critical Analysis* (London: Routledge and Kegan Paul, 1958), 81–2.
95. Ellul, *The Presence of the Kingdom*, 80.
96. Anders, *Die Antiquiertheit des Menschen*, vol. 2, 239.
97. Ibid., 239.
98. Ellul, *Propaganda*, 124.
99. Villa, *Public Freedom*, 170.
100. Ibid.
101. Ibid.

6

TECHNOLOGIES OF THE BODY: MAN AS RAW MATERIAL

INTRODUCTION

'This is the new era of "Participatory Evolution"', wrote David Rorvik in 1971. 'No longer is man the offspring of Nature, the creature of natural selection. Science has provided him with the technology to become his *own* maker.'[1] Rorvik was referring to the new science of genetics, but technology was seen to have made man into his own maker in many different ways. The body is at the centre of technology's transformation of modernity. Each of the technological processes that the preceding chapters outline can be described in terms of the body: the destructive power of technology imposes its brutality upon the body; technology's influence on labour transforms the way that the body operates and is worked upon within the process of production (and, by extension, in leisure and consumption activities); mass culture driven by media technologies organises the leisure activities of the people as a unified body while propaganda manipulates the actions of the body and bodies of the people through its own means. The body, as the material centre of the self, is involved in every technological change that takes place. It thus appears at the intersection of science and technology: biological warfare, genetic manipulation, atomic fallout. Considering the body specifically, however, and the way that the critics of technology believe it is being uniquely manipulated by modern technology, more clearly reveals their distinct claims about what matters about human beings, and what the relationship between nature and technology is, and is becoming.

163

In twentieth-century political thought, Michel Foucault is perhaps the most famous thinker to explore the body: the way that the body is worked upon by social and political elements of power, and the way that treatment of the body – in prisons, asylums and other arenas – reveals the form of the political and its transformations. His work highlights the centrality of the body in modern political systems, and the political quality of the body itself. In work from the 1970s onwards, Foucault describes the 'technology of the self' (in *The History of Sexuality*) and 'political technologies of the body' (in *Discipline and Punish*). The body, he argues, is always the subject of social and political ways of ordering: the organisation of the body, the way in which it is understood, and the ways in which it is physically treated and managed, comprises its own 'technology'. Thus technology is used by Foucault in a more expansive sense than that used by the critics of technology: it is a way of describing the manner in which bodies are worked on and through by the prevailing social and political forces, rather than a reference to a specifically modern form of organisation and production, possessing its own political momentum. Although Foucault was influenced by the anti-technological mood in past-war France in the very early part of his career, his idea of technology ultimately took a different direction from the critics of technology.[2]

Despite overlapping influences – the French context of anti-technologism in the 1950s and, notably, the influence of Heidegger on Foucault's thinking – Foucault cannot be considered part of the critical discourse on technology outlined in this work, not least because his idea of technology is significantly more neutral, or, at least, less catastrophic, than those of the other theorists explored here. While he describes historical changes in the way that the body is worked upon and utilised by social and political forces, it is not the case that these changes represent a problem that is distinctly modern. The body has *always* been the locus of authority; it has always been manipulated and utilised in order to achieve particular ends or to convey particular messages. One might disagree with this, or even consider it an illegitimate form of power, but it is not distinct to modernity, nor can modern technology be considered *responsible*. Foucault's conceptualisation of technologies of the body also differs from the way technology is understood by the technological sceptics in one further, related, way: technology does not appear as something that is fundamentally alien to humans, but rather technology describes processes and practices that are part of normal human functioning (albeit normal human functioning is always subject to social organisation and manipulation). The way in which Foucault conceptualises 'the body' in relation to technology is thus quite different from the critics of technology, as this chapter will show.

Although the body might be considered to be at the centre of the relationship between technology and politics, there is substantial variation in how extensively the critics of technology deal with the body in their work on this theme.

For instance, Heidegger does not discuss the body in any significant way in his analyses of technology, or of Being – the body is not considered part *of* Being – as Jonas takes particular exception to. Adorno and Horkheimer's thinking on technology in modernity prioritises understanding rationalisation and reason, the influence of technology on the mind, rather than the body. Yet, in one note in *Dialectic of Enlightenment*, they connect the practice of technological domination with the diminished body when they write that 'in man's denigration of his own body, nature takes its revenge for the fact that it has reduced nature to an object for domination, a raw material'.[3] The transformation of the body into a living corpse, they later add, 'was simply a part of that perennial process which turned nature into substance and matter'.[4] The status of the body reflects what contemporary humans have imposed upon nature itself.

Others – Ellul, Mumford, Marcuse and Anders – do not centre discussion of the body in their work, but at times touch upon the subject. For two theorists here, however, the body – discussed in terms of biological life, or the human organism – features prominently in their conceptualisation of contemporary technology: Arendt and Jonas. After the war, writes Jonas, 'to the fear of a sudden [nuclear] catastrophe was soon added the growing realization of the negative side of technological triumphs in general'.[5] On the questions that sprang out of new advances in biology and medicine, he argues

> there is no longer room here for a simple 'yes' or 'no' as with the problem of nuclear weapons; instead, we find an area of fluid boundaries, subtle value judgments, and controversial decisions . . . Biotechnology in particular has introduced into the realm of morality completely new dilemmas, heightened complications, and refined nuances.[6]

Biotechnology reveals technology's 'previously undreamt-of power', he concludes, and the 'previously undreamt-of challenges' that accompany it.[7]

Jonas's and Arendt's ideas on the relationship of human biology to politics are in many ways at odds. Yet, this chapter will make the claim that there are, nonetheless, important similarities in how they understand the treatment of the body in the technological era, and make the case that the changes they identify are also echoed in the work of the other theorists, at points. Both Jonas and Arendt, in different ways, make the case that modernity is transforming the body into nothing more than raw material, a phraseology that can also be found in the work of Anders, Heidegger, Adorno and Horkheimer. The incessantly rationalising force of technology thus desacralises even the physical body, in a process that, although *preceding* technological modernity, is accelerated and accentuated by the development of technology. The rationalisation and desacralisation of the body operates at the very deepest level: it is all-consuming, existentially nihilistic, because it strips away anything that is not 'necessary' from the body; the body is no longer (part of) the person, the end in itself, but merely means, or material. Bodies, as raw material, become something that technology works upon, uses or even consumes, rather than objects imbued with value, meaningfulness or sacredness. Such bodies are also deindividualised, that is, they become part of a wider, amorphous mass, considered to exhibit 'behaviour' rather than free action. Technology thus poses new and particular threats to the body by increasingly reducing it to a *mere* body, and it changes the way we think about and see the body by framing it as a rationalised, secularised object, which is treated as such by technological practices. Technology thus seeks to control the body, to shape it, manipulate it and, often, to destroy it, and thus shapes the political in the most fundamental way.

The Manufacture of Corpses: The Body as Raw Material

Hannah Arendt's analysis of concentration camps in *The Origins of Totalitarianism* outlines the methods by which their inhabitants were reduced to something less than human. Her explanation of the camps, and the way in which Nazi Germany progressively dehumanised those who eventually came to live and die within them, is generally framed in terms of Arendt's concept of the political: that the 'rights' of man mean nothing without a political community willing to defend such rights – that the notion of 'human rights' is nonsensical; that there are only political, constructed rights (if rights are to have any force, or any true meaning at all). For Arendt, the 'total humanization of reality [is viewed] as the most extreme form of alienation', as Villa writes.[8] This argument is at the heart of Arendt's explanation of the *success* of Nazi fascism: that contemporary political understanding is fatally flawed by its failure to comprehend the artificial, unnatural quality of the political. Where individuals do not 'belong' to a political community, that is, are not claimed or recognised by any community, or where they can be deemed to be excluded from that community, the conditions are in place for their reduction to something less than criminal, less than human, to a mere body. The decline of the person to nothing more than a body – and ultimately a corpse – is also present in the work of other critics of technology in their considerations of the death camps. 'The transformation of man into raw materials', writes Günther Anders, began in Auschwitz, where 'people weren't killed, rather, corpses were manufactured.'[9] However, another component is also present in Arendt's understanding of the way in which humans are reduced to mere life, which can be traced in *The Origins of Totalitarianism*, and later in *The Human Condition* where she takes up the themes of human rights and political community. That is the way in which labour – biological life – increasingly becomes prioritised in modernity, along with the absolute prioritisation of life itself as the absolute good, in a secular transformation of Christian ideology.

In the concentration camps, Arendt wrote, the 'mass manufacture of corpses' was achieved.[10] This 'insane' process was preceded, however, by the 'historically and politically intelligible preparation of living corpses', the reduction of the person to a mere body.[11] She outlines the historical process by which this took place in Nazi Germany, through which the Jews were cast out of the society of rights. Their first loss was of their homes 'and this meant the loss of the entire social texture into which they were born and in which they established for themselves a distinct place in the world'.[12] The second loss was of government protection, 'and this did not imply just the loss of legal status in their own, but in all countries'.[13]

> Only in the last stage of a rather lengthy process is their right to live threatened; only if they remain perfectly 'superfluous,' if nobody can be found to 'claim' them, may their lives be in danger. Even if the Nazis started their extermination of the Jews by first depriving them of all legal status (the status of second-class citizenship) and cutting them off from the world of the living by herding them into ghettos and concentration camps; and before they set the gas chambers into motion they had carefully tested the ground and found to their satisfaction that no country would claim these people. The point is that a condition of complete rightlessness was created before the right to live was challenged.[14]

The condition of complete rightlessness led Arendt to assert the necessity of the 'right to have rights': to be considered part of the *human* community, or to have one's 'human rights' guaranteed, the membership of a polity was a necessary prelude. Such a right, Arendt argued, was not and could not be expressed in the categories of the eighteenth century, and the emergence of rights-discourse, because they presumed that 'rights spring immediately from the "nature" of man – whereby it makes relatively little difference whether this nature is visualised in terms of the natural law or in terms of a being created in the image of God, whether it concerns "natural" rights or "divine" commands'.[15] The political and thus artificial quality of rights that Arendt asserts is necessary is the only thing that can ensure the dignity of humans in practice. Neither a person's 'nature' nor their supposed sacredness can defend against their enforced exclusion from the community of human beings, or indeed their annihilation, as the death camps revealed.

The rights of man depended upon 'the absolute and transcendent measurements of religion or the law of nature', writes Arendt.[16] Once these have lost their authority, as she argues they have in the modern world, 'a conception of law which identifies what is right with the notion of what is good – for the individual, or the family, or the people, or the largest number – becomes inevitable.'[17] As such, the concept of inalienable rights is replaced by a calculation of benefit:

> it is quite conceivable and even within the realm of practical political possibilities, that one fine day a highly organized and mechanized humanity

> will conclude quite democratically – namely by majority decision – that for humanity as a whole it would be better to liquidate certain parts thereof.[18]

In the twentieth century, man has become as 'emancipated from nature as eighteenth-century man was from history', writes Arendt.[19] The laws of nature masked the problems of political philosophy, just as Christian theology enabled 'the oldest perplexities of political philosophy ... [to] remain undetected'.[20] Both the concept of Christianity and the concept of nature held some politically stabilising social value therefore. Yet the latter is also double-sided. The *laws* of nature held some capacity to uphold norms, yet nature itself bears no such ability, and in fact, when people are excluded from citizenship and thrown back upon nature, or their 'natural givenness', they come to lack 'that tremendous equalizing of differences which comes from being citizens of some common-wealth'. In this situation, 'since they are no longer allowed to participate in the human artifice, they begin to belong to the human race in much the same way as animals belong to specific animal species'.[21] Where once the laws of nature endowed humans with value, the reality of humans thrown back upon their natural existence is to strip them of their specifically human value. The 'laws of nature' therefore much more closely resemble the ideological system Arendt tends to pair with them in her discussion of historical rights – that of Christian theology – than the reality of nature itself. In nature, we are an 'animal species', not political animals.

The manufacture of corpses that took place in the death camps revealed the poverty of Western man's rationalised conception of man as a 'natural' entity, and the reduction of humans to mere bodies. In a context where traditional belief systems had lost their credibility and authority, rights, dignity and eventually life itself were easily stripped from millions of individuals. The concentration camps, and the instruments of death, are clearly forms of technology: the camps themselves forms of bureaucratically ordered organisation, relying upon modern technology in various ways; the novel means of execution in the gas chambers were themselves integral parts of the process of annihilation. But the explicit relationship between technology and bodies in the concentration camps is only developed in Arendt's later work, *The Human Condition*, where she deals with the topic of technology in modernity more comprehensively, finding the camps to be a (perhaps inevitable) consequence of technologised modernity.

It is the human world, the artifice of the political community, that separates human existence from that of animals, argued Arendt. Technology has for a long time been engaged in an attempted transformation of the human condition, an attempt to escape its limitations: 'A great many scientific endeavors have been directed toward making life also "artificial," toward cutting the last tie through which even man belongs among the children of nature.' The desire to 'escape from imprisonment to the earth' can be found in our attempt to create life in laboratories, or in our hopes of extending the human lifespan beyond its natural limits, she claims.[22] Whether we should use our new technological and scientific powers in this way is a 'political question of the first order', she notes, and she herself clearly believes the attempt to transform life itself should be left well alone.[23] What we are increasingly doing is stripping the human being of everything that gives it meaning. We are, Arendt claims, increasingly defined in terms of life itself or life alone: bodily existence. We define ourselves now as 'universal', she claims; 'creatures who are terrestrial not by nature and essence but only by the condition of being alive.'[24]

Alongside this, the development of automated technology increasingly means that the rhythm of machines replaces the rhythm of the human body. A human that is *just* a body (without civil, social or political protections) is enfeebled, and thus at the mercy of these demands. 'The machines demand that the laborer serve them ... Even the most primitive machine guides the body's labor and eventually replaces it altogether.'[25] Our technology increasingly comes to look like part of the biological process itself, not under the control of human beings, but increasingly shaping, manipulating and utilising the bodily form of the human. The 'world of machines' feeds on 'natural processes', Arendt argues, and these 'increasingly relate it to the biological process itself, so that the apparatuses we once handled freely begin to look as though they were "shells belonging to the human body as the shell belongs to the body of a turtle"'.[26]

A human body, outside the protective environment of the political community and the rights and protections afforded by that artifice, is therefore a fragile and insubstantial object, prone to bend to the machine's demands. When humans are considered as 'natural' entities, they are little else. In this context, technology comes to overlay the natural, the relationship between the human and the machine is flipped from the historical tool-oriented technological form, where humans used tools for their own productive purposes, to a condition where the body is increasingly at the mercy of, and used for the progress of, the machines themselves.

Although technology has come to be a transformational force in almost every sphere of human life, it is not the only transformation that has taken place in modernity that has shaped how the body is understood in the contemporary world. Arendt identifies a peculiarity in modern thought: the combining of secularism with a continued belief in certain Christian dogmas. Today, she writes, 'life itself' is considered the highest good:

> The reason why life asserted itself as the ultimate point of reference in the modern age and has remained the highest good of modern society is

that the modern reversal operated within the fabric of a Christian society whose fundamental belief in the sacredness of life has survived, and has even remained completely unshaken by secularization and the general decline of the Christian faith.[27]

The way in which the Christian church prioritised 'life itself', Arendt argues, reversed 'the ancient relationship between man and world and promoted the most mortal thing, human life, to the position of immortality, which up to then the cosmos had held'. Individual life, Arendt writes, 'now came to occupy the position once held by the "life" of the body politic'.[28] In the contemporary secular world, while the Christian faith in the immortality of life is disappearing, the belief in the prioritisation of life in and of itself remains. This has implications for the political character of modernity, which retains some of the characteristics of the Christian view of politics, in a context where the Christian emphasis on the contemplative life has given way to a resurgence of the active life, and the importance of the temporal world – the resurgence of the political (in some ways). The Christian emphasis on the sacredness of life 'tended to view labor, work and action as equally subject to the necessity of present life' – in worldly terms, there was a somewhat utilitarian emphasis on the necessary rather than the higher political purposes of existence:[29]

> the modern age continued to operate under the assumption that life, and not the world, is the highest good of man; in its boldest and most radical revisions and criticisms of traditional beliefs and concepts, it never even thought of challenging this fundamental reversal which Christianity had brought into the dying ancient world . . . what matters today is not the immortality of life, but that life is the highest good.[30]

What has a secularised faith in 'life itself' to do with the atrocities of the concentration camps? The problem, for Arendt, is that by rooting the concept of human meaningfulness in 'individual life' or 'life itself', the political is sacrificed for the biological. It is biological life, the body or metabolism, that is now prioritised, along with those actions necessary for bare survival (labour). Darwinianism was a 'turning point in the intellectual history of the modern age', she writes, because organic life development replaced the earlier 'mechanistic world view'.[31] In a post-Cartesian world, it found 'within man – not in his mind but in his bodily processes – enough outside matter to connect him again with the outer world'.[32] The emphasis on 'life itself' does not enforce protections for individual lives, but indicates a faith in biological life. Only within the political world can individuals be recognised as individuals; only by virtue of belonging to a political community of equals can the individual's life be guaranteed. The modern emphasis on biological life above political existence thus becomes an enabling factor in the emergence of the death camps.

Thus we move from the Christian sanctity of life itself, as the (temporal) origin of the immortal soul, to a post-Christian regard for life itself understood not in terms of the individual soul, but rather as the biological life of the human animal or species, whereby the dignity and rights of the individual are, against the 'good of the whole', disregarded. Contemporary science even attempts to move towards the construction or creation of life itself, doing 'what all ages before ours thought to be the exclusive prerogative of divine action'.[33] Nonetheless, she points out, this technological ambition to harness and create life, 'through blasphemous in every traditional Western or Eastern philosophic or theological frame of reference . . . is no more blasphemous than what we have been doing and what we are aspiring to do'.[34] The contemporary technological world stands in contrast to even the mechanised technological world of the industrial era:

> Industrialization still consisted primarily of the mechanization of work processes, the improvement in the making of objects, and man's attitude to nature still remained that of *homo faber*, to whom nature gives the material out of which the human artifice is erected.[35]

Today's technology is shaped rather by the desire of modern man to act into nature, 'creating natural processes and directing them into the human artifice and the realm of human affairs'.[36]

Arendt's analysis of the process by which Jews and other persecuted people were progressively stripped of their rights, in such a way that they lost all protections, and became completely vulnerable, certainly rests upon the claim that only membership of a political world or community can guarantee such rights. But Arendt details a longer history of this process, which is the history of the evolution of the concept of 'life'. With the rise of Christianity and faith in an immortal sphere, a decline in the importance of the temporal or political world took place. The Christian concept of life, which begins in this world, and continues into the next, released man from the chains of mortality (something which had formerly only been possible, albeit in a very different sense, through one's actions in the construction of a political world extending into the future, beyond the limits of individual life). 'Life itself' in Christian terms refers to the sanctity of the individual soul. In modern secular terms, however, life itself comes to refer to the biological process, not the life of the individual. To 'manufacture corpses' as occurred in the camps is therefore not, astonishingly, against the strictures of the contemporary faith in life itself; if it can be justified by some (however spurious) appeal to the greater good.

Technology only becomes part of this long story in its late stages: technology certainly did not initiate the transformation of the concept of life, nor was it responsible for the shift from person to body as biological organism or species. But in its various guises, whether concerned with creation or annihilation, technology increasingly treats bodies as mere raw material for its ends. Technology becomes a kind of 'natural' process, the biological itself, technology sees life through the lens of biology, not individuality. The body of the individual, therefore, is no longer sacred in the secular, technological world, nor is it necessarily protected by a political world, but is an object to be worked upon, used or consumed. It reduces humans to something less than human, and no more meaningful than any other worldly object.

CREATING AND RECREATING THE BODY: BIOTECHNOLOGY IN THE LATE TWENTIETH CENTURY

Hans Jonas's concern with technology's influence on the body took a different focus: the influence of the nascent field of biological engineering. Until now, he writes, technology has been concerned with creating human artefacts out of lifeless materials. 'The advent of biological engineering signals a radical departure from this clear division,' he argues, 'indeed a break of metaphysical importance: Man becomes the direct object as well as the subject of the engineering art.'[37] This is a literal and radical reshaping of the body. Man has assumed increasingly more sovereign roles, he writes, 'involving a deeper meddling with the patterns of nature – indeed a redesigning of such patterns . . . Man steps into nature's shoes, and from utilizing and exploiting he advances to creating.'[38] The possibility of self-limits on the use of novel technologies is unlikely, argues Jonas; 'the idea of taking evolution into our own hands is intoxicating even to men of science, who should know better.'[39] It is all the more unlikely because these technologies are being used not for evil but for good; to benefit humanity in myriad ways. This is the nature of the danger, Jonas insists. 'Speaking for myself, I fear not the abuses by evil power interests; I fear the well-wishers of mankind with their dreams of a glorious improvement of the race.'[40]

Biological technologies promise to change the very fabric of human existence: perhaps indefinitely extending possible life spans, such that 'death no longer appears as a necessity belonging to the nature of life, but as an avoidable, at least in principle tractable and long-delayable, organic malfunction'; or by developing forms of behaviour control, a possibility 'nearer to practical readiness', Jonas believed.[41] These pose profound ethical challenges of different kinds, which is the focus of much of Jonas's work on the subject. In our technological endeavours to recreate man's physical nature, or to take control of human evolution, we *must* determine what is or should be the image of man that should guide us, he urges. 'Who will be the image-makers, by what standards, and on the basis of what knowledge?'[42] This task becomes urgent because of the new challenges of technology and because, like Arendt,

172

he makes the case that the human organism has been increasingly stripped of its meaningfulness, or 'neutralized'.[43]

It is not just the promise of technology that challenges contemporary ethics and our understanding of human nature. Jonas's critique, in a 1969 essay, of a proposed redefinition of death to include irreversible coma highlights how our understanding of mortality is already being pressed by advances in medical technology. It also indicates how humans are being treated increasingly as *mere* bodies, or at least that there is technologically driven pressure to do so.

> I see lurking behind the proposed definition of death, apart from its obvious pragmatic motivation, a curious revenant of the old soul-body dualism. Its new apparition is the dualism of brain and body. In a certain analogy to the former it holds that the true human person rests in (or is represented by) the brain, of which the rest of the body is a mere subservient tool. Thus, when the brain dies, it is as when the soul departed: what are left are 'mortal remains.'[44]

While not denying the importance of the brain to the human person, he writes, 'it is no less an exaggeration of the cerebral aspect as it was of the conscious soul to deny the extracerebral body its essential share in the identity of the person.' The identity of a person encompasses the *whole* organism, Jonas insists: 'The body is as uniquely the body of this brain and no other, as the brain is uniquely the brain of this body and no other.'[45] He views this as the only appropriate stance for phenomenology, correcting what he perceived to be Heidegger's association of being with mind, and directly against Cartesian rationalism and the Cartesian view of the 'animal machine' that he argues is the only position that could justify a claim that any *still-living body* (even one without a functional or conscious mind) was actually dead.[46] On this basis, he makes the case that 'the body of the comatose, so long as – even with the help of art – it still breathes, pulses, and functions otherwise, must still be considered a residual continuance of the subject that loved and was loved'. [47] Against the perceived diminishment of the body to *mere* body, a non-person, by the writers of the new definition of medical death, the body is 'still entitled to some of the sacrosanctity accorded to such a subject by the laws of God and men. That sacrosanctity decrees that it must not be used as a mere means.'[48]

The argument that Jonas makes in this article is that individuals ought not to be kept alive by artificial means solely in order to maintain the quality of organs for transplant (into other bodies); that the dignity and sanctity of the individual must be prioritised above such questions of utility, however principled the aim. 'The cowardice of modern secular society which shrinks from death as an unmitigated evil needs the assurance (or fiction) that he is already dead when the decision is to be made,' he writes.[49]

> Insofar as the redefiners of death – by saying 'he is already dead' – seek to allay the scruples about turning the respirator off they cater to this modern cowardice which has forgotten that death has its own fitness and dignity, and that a man has a right to be let die . . . they serve the ruling pragmatism of our times which will let no ancient fear and trembling interfere with the relentless expanding of the realm of sheer thinghood and unrestricted utility. The 'splendor and misery' of our age dwells in that irresistible tide.[50]

Current and prospective technological developments in the field of medicine and biological engineering are, for Jonas, clearly challenging our ideas of what it is to be human. But these very modern issues emerge from a view of the human organism that, despite having been dominant for hundreds of years, is basically wrong, according to Jonas. Philosophy recognises the person only as mind, he argues, 'but the idealism of consciousness – whether in the form of neo-Kantianism, phenomenology, or existentialism . . . exposed no more than the tip of the iceberg of our being and left submerged the broad organic basis on which the miracle of mind is perched'.[51] This derives from Cartesian dualism, he argues, 'which has split reality into *res cogitans* and *res extensa*'.[52] Against a philosophy that has given up the field of nature to science, Jonas argues, 'the *organism* with its insoluble fusion of inwardness and outwardness constituted the crucial counterevidence to the dualistic division . . . indeed the key to a reintegration of fragmented ontology into a uniform theory of being'.[53]

'In the body,' Jonas writes, 'the knot of being is tied which dualism does not unravel, but cut.'[54] It is thus to the body, or more specifically, the *organism* that we must return to understand humans in their entirely, not simply as consciousness, mind, reason or, more recently, brain. Descartes offered a mechanistic theory of the animal body, Jonas, explains, an 'animal automata . . . animals, in other words are nothing but bodies'.[55] Against this, man became 'an inexplicable, extraneous combination of mind and body – a combination with no intelligible relevance of the body for existence and inner life of the mind (as also, of course, vice versa)'.[56] Not only was Descartes's image of man radically incomplete, it also set the philosophical conditions by which the distinction of man within the animal kingdom could be nullified, whereby 'the human body became no less an automaton than all other organisms'.[57]

Against this, Jonas argues for a reintegrated concept of the human being, as both mind and body, literally inseparable but also philosophically incomprehensible in separation. Man is not mind alone but a certain kind of metabolism, that is, an integration of body and mind, and the form this takes is *self-conscious* metabolism – the body is absolutely part of the essence of man, and inseparable from it. But consciousness, or self-consciousness, while it transforms man into a different kind of animal, does not mean that he transcends and leaves behind his organic self, simply that he becomes a more complex kind of natural being, namely (among other things) an entity that has the capacity for ethical action, and the consequential responsibility to act in ways that are ethically sound. Through an understanding of the human being in its wholeness, Jonas attempts to construct a philosophy that might stand up against what he considers to be the nihilism in contemporary philosophies of being that consider man as nothing *but* consciousness, namely, of course, that of his former teacher Heidegger.

One way in which the limits and problems entailed by the Cartesian image of the divided individual are very clearly visible is in the field of contemporary biological engineering. Where bodies are increasingly seen as 'automata', new technological capacities enable scientists to turn to the creation and reshaping of humans and humanity itself. The late 1960s in particular, he writes, 'saw the emergence of the most acute, internal crisis of the Baconian ideal, compared with which all former, external challenges by romantic, traditionalists, or malcontents had been child's play'.[58] The extent of the power that was granted to humanity by virtue of its new technologies meant traditional norms and modes of understanding were increasingly irrelevant. Among modern technology's extraordinary powers for manipulation he lists 'the automation of all work, psychological and biological behavior control, various forms of totalitarianism, and – probably most dangerous of all – the genetic reshaping of our nature'.[59] While each of these entail ethical challenges, the potential for the manipulation of man's physical nature places humanity in a new category. 'Man himself has been added to the objects of technology. *Homo faber* is turning upon himself and gets ready to make over the maker of all the rest.'[60]

Ethical powerlessness, an incapacity to deal with the ethical questions which characterise our age, and a fundamentally nihilistic approach to existence thus characterises our era. Like Arendt, Jonas emphasises the role secularisation has played. Part of this is due to the triumph of reason, through science, which has 'destroyed the faith in revelation, without, however replacing revelation in the office of guiding our ultimate choices'.[61] But the nihilistic perspective of modern man is due to more than just the triumph of reason, and Jonas points to 'the modern concept of nature, the modern concept of man, and the fact of modern technology supported by both. All three imply the negation of certain fundamental tenets of the philosophical as well as the religious tradition.'[62]

Against biblical concepts, according to modern science, the world is not created but has made and is making itself, writes Jonas.[63] This view of nature is reflected in the modern concept of man: 'In the Darwinian view, man bears no eternal "image" but is part of the universal, and in particular of biological "becoming" . . . He is an accident, sanctioned merely by success.'[64] Psychological explanations of man's nature reinforce this, Jonas argues, by 'cutting man down to size and stripping him in his own eyes of every vestige of metaphysical dignity . . . it has become the common currency of our everyday psychologizing: the higher in man is a disguised form of the lower.'[65] Humans, like nature itself, are nothing more than an accident; man's nature has been essentially diminished, and traditional norms offer no answers to how humanity ought to be treated. 'It is the tremendous *power* which modern *technology* puts into his hands . . . a power therefore which has to be exercised in a vacuum of norms, that creates the main problem for contemporary ethics.'[66]

> With his very triumph, man has become engulfed in the metaphysical devaluation which was the premise and the consequence of that triumph. For he must see himself as part of that nature which he has found to be manipulable and which he learns how to manipulate more and more . . . Man today, or very soon, can make man 'to specification' – today already through socio-political and psychological techniques, tomorrow through biological engineering, eventually perhaps through the juggling of genes. The last prospect is the most terrifying of all. Against this power of his own, man is as unprotected by an inviolable principle of ultimate metaphysical integrity as external nature is in its subjection to his desires: those desires themselves he may now undertake to 'program' in advice – according to what?[67]

These changes in the understanding of the body and the nature of the self pose not only ethical questions and challenges, but also political ones. In *The Imperative of Responsibility*, Jonas explains how the unpredictability and heightened degree of risk that technology introduces into the modern world transforms previously ethical questions into political ones, simply because the scope of human action, through technology, is vastly greater and potentially more threatening than has ever been the case before. But another kind of politicisation can also be seen to be taking place, of the kind which Foucault and later Agamben also come to write about: the way in which power comes to be exercised through the manipulation of the body, and the idea of what it is to be 'human' itself. As Agamben wrote, for the Nazis, 'the Jew is a human being who has been deprived of all *Würde*, all dignity: he is merely human – and, for this reason, non-human.'[68] Today, Agamben argues, 'modern politics is, from top to bottom, biopolitics'.[69] For Agamben, this relates particularly to medicine, and its object, the human body. Here 'a subtle and rigorous dogmatics corresponds, in praxis, [to] a vast and intricate cultic sphere that coincides with what we call "technology"'.[70] When, as Jonas believes has occurred, the human body is stripped of its traditional meaning and dignity – when it has been desacralised – it opens up the possibility for the body to be politicised and manipulated in novel and dangerous ways.

MASS SOCIETY AND THE BODY: BEHAVIOURISM AND SOCIAL UNIFORMITY

Despite the progress of technology in modernity, and the unprecedented power this has placed in human hands, according to Arendt and Jonas, individuals within technological societies are increasingly inclined to *behave* in ways that exhibit powerlessness, that is, to fall into line with patterns of behaviour.

'Economics', wrote Arendt, 'could achieve a scientific character only when men had become social beings and unanimously followed certain patterns of behavior, so that those who did not keep the rules could be considered to be asocial or abnormal.'[71] A critic of behaviourism, she nonetheless notes that 'the unfortunate truth about behaviorism and the validity of its "laws" is that the more people there are, the more likely they are to behave and the less likely to tolerate non-behavior'.[72] This leads to the 'levelling out' of difference, and a statistical uniformity which is the 'political ideal of a society which, entirely submerged in the routine of everyday living, is at peace with the scientific outlook inherent in its very existence'.[73]

The development of the science of economics was followed, she writes, by the rise of the social sciences as '"behavioral sciences," [which] aim to reduce man as a whole, in all his activities, to the level of a conditioned and behaving animal', a stage which reflects the dominance of mass society and 'social behaviour' as the standard for action.[74] The growth of 'social behaviour' or the subsumption of politics by 'the social', which she writes has been accelerating for 300 years, 'derives its strength from the fact that through society it is the life process itself which in one form or another has been channeled into the public realm'.[75] Antiquity was contemptuous of the private, the space of the life process, precisely because 'man existed in this sphere not as a truly human being but only as a specimen of the animal species man-kind'.[76] The same, Arendt suggests, is true today, but where life itself, the biological process, and the private sphere have become public and political, the individual can be nothing more than a human animal. 'The trouble with modern theories of behaviorism is not that they are wrong,' Arendt writes, 'but that they could become true, that they actually are the best possible conceptualization of certain obvious trends in modern society.'[77]

In *The Phenomenon of Life*, Jonas offers a critique of the cybernetic theory of Norbert Weiner, which seeks to outline general principles of human behaviour through mathematical analysis. 'Scientists, for so long the very abjurors of anthropomorphism as the sin of sins, are now the most liberal in endowing machines with manlike features,' Jonas argues. This irony, he writes, is 'only dimmed by the fact that the real intent of the liberality is to appropriate the donor, man, all the more securely to the realm of machine'.[78] The cybernetic model falls short, he argues, in failing to recognise that 'living things are creatures of need . . . based both on the necessity for the continuous self-renewal of the organism by the metabolic process, and on the organism's elemental urge thus precariously to continue itself'. This, he claims, 'manifests itself on the level of animality as appetite, fear and all the rest of the emotions'.[79] Man possesses these animal features, but also 'alone in the world thinks, not because but in spite of his being part of nature. As he shares no longer in a meaning of nature, but merely, through his body, in its mechanical determination, so nature no longer shares in his inner concerns . . . This is the human condition.'[80] The uniqueness of humans is that they exist both in and outside nature, are both body and mind. Any model that considers only one part of the dual nature of man, is bound to misunderstand the nature of humanity.

These ideas are reflected in the work of other critics of technology. 'The awareness of the self as an autonomous individual with his own soul is giving way to the corporate mentality,' wrote Horkheimer.[81] 'The meaning which accrued to every action in life from the thought of eternity is being replaced by an absolutizing of the collective, into which the individual feels himself incorporated.'[82] Through the matrix of modern production, consumers are bound 'more or less pleasantly to the producers, and through the latter, to the whole', Marcuse explains.[83] 'The machine process in the technological universe breaks the innermost privacy of freedom and joins sexuality and labor in one unconscious, rhythmic automatism – a process which parallels the assimilation of jobs.'[84] Such is 'the pure form of servitude: to exist as an instrument, as a thing'.[85]

One space in which several of the theorists of technology identify the manipulation of the body as a political act is that of sport. Adorno, Horkheimer, Anders and Ellul all express concerns with the relationship between technology and behaviour that they believe is revealed by the role of sport in modern societies. Sport is an exemplar of how bodies are used and shaped by totalitarian forces at work in modernity. Their characterisations are clearly influenced by the Nazi idealisation of sport culture and physical appearance, or the political significance of sport for the Soviet Union. But as a mode of technique, this use of sport is far more widespread, they argue.

'Sport is linked with the technical world because sport itself is a technique,' writes Ellul.[86] 'For the Greeks, physical exercise was an ethic for developing freely and harmoniously the form and strength of the human body. For the Romans, it was a technique for increasing the legionnaire's efficiency. The Roman conception prevails today.'[87] In precision training and discipline, 'we find repeated in sport one of the essential elements of industrial life . . . the human being becomes a kind of machine'.[88] The individual is disciplined through this sport, finding 'relaxation from the various compulsions to which he is subjected, but without knowing it trains himself for new compulsions . . . improvisation and spontaneity all disappear'.[89] The political aim of sports today, he argues, is not the training of elite athletes, but rather 'the extension of the sporting mentality to the masses'. It results in the integration of 'more and more innocents into an insidious technique'. The exercise of sport, he concludes, paves the way for 'the totalitarian frame of mind'.[90] 'Technicized sport' was first developed by the United States, writes Ellul, 'the most conformist of all countries'. It was only *later* developed by dictatorships 'to the point that it became an indispensable constituent element of totalitarian regimes'.[91] It was an 'essential factor' in the creation of mass society, he argues; it 'coincides exactly with totalitarian and with technical culture . . . In every conceivable way sport is an extension of the technical spirit.'[92] The body *itself* becomes merely a functionary of technique. Anders makes similar observations. Like leisure and consumption, 'sports are imposed on us just like work', an 'unfreedom'.[93] The experiences of the sports stadium are of interest in the same way as the processes of assembly line labour or automated labor processes, he writes.[94] Adorno and Horkheimer likewise argue that 'athletic barbarity completes what has threatened the creations of the spirit since they were gathered together as culture and neutralised'.[95]

THE DESACRALISATION OF THE BODY: TURNING MEN INTO THINGS

Arendt's and Jonas's ideas are opposed in some important respects. In part, this comes from the fact that Arendt is a political theorist, concerned with exploring the space of the political and the effects of modernity on that space, while Jonas's focus is very clearly on the ethical challenges posed by modernity. Those areas overlap, but also mean that the two are working from different positions from the start.

Most importantly, Arendt considers the contemporary shift towards the prioritisation of the natural, biological sphere or category to be fundamentally antipolitical, because it diminishes the distinctiveness and priority of the political sphere, through which rights have meaning. For Arendt, the treatment of the body as mere material comes about because of the even more significant problem of the wider depoliticisation she describes. For Jonas, it is the diminishment of the biological element of human beings that has resulted in the nihilistic attitude towards human beings as merely disembodied mind. For Jonas, the body's relationship with the human in the broadest sense *must* be taken seriously; the human cannot be understood only in a partial sense as mind. This is a fundamental clash of opinions. Arendt's argument, for example, that nineteenth-century naturalism found the key to man in man's own biology (and her critique of what, for her, is a basic misunderstanding about what *matters* about man) is the exact opposite of Jonas's claim that the meaning of human existence – our freedom, the responsibilities we bear, and the character of humans – derives from our nature as self-conscious metabolism.[96]

As Jonas wrote in his memoirs, Arendt objected strongly to the ideas he offered in *The Imperative of Responsibility*;

> She absolutely rejected my argument that man's fundamental responsibility could be explained biologically, on the basis of the natural order. In her view, responsibility was a voluntary relationship derived from the polis, from civic or political life . . . [she] was of the opinion that responsibility for the general welfare was essentially artificial and unnatural, a product of the *contrat social*, according to Western tradition. We agreed that modern technology was becoming a threat to the world and that we humans bore responsibility for the future. But her philosophical justification for this position would have looked very different from mine.[97]

Of the two perspectives, Jonas's is the more widely endorsed among the other theorists of technology. Mumford, for instance, draws heavily in his work on the distinction between organicism and automatism (arguing for a return to the former). Automation, for him, is the *opposite* of what Arendt claims, that it is the realisation of biological processes. The effect of 'the machine and machine-made men . . . denaturing or banishing every organic attribute', he writes, has been to create 'an environment . . . fit only for machines'.[98] Marcuse insists that liberation must involve 'the mind and the body . . . [the] entire human existence.'[99] Socialism, the total social revolution Marcuse agitates for, is a biological, sociological and political necessity, he argues: a biological necessity in as much as a socialist society, according to Marx, would conform with the very *logos* of life, with the essential possibilities of a human existence, not only mentally, not only intellectually, but also organically.[100]

But despite the differences between Arendt and Jonas, both make a similar argument about the relationship between the body and technology in the contemporary world. The body – the individual body – is increasingly stripped of meaning and devalued. This means the body can be treated as a mere object or as raw material, and modern technologies increasingly utilise man in this way, extending the objectifying process that began with secularisation. Men are being turned into material through technologisation. Against this, Jonas urges, our 'protest should always be against turning men into things'.[101] Anders expresses an even more extreme reading: not only are humans material, but are considered saboteurs of technology; 'they are too emphatically defined to keep up with the daily changing world of machines; a world which makes a mockery of all self-determination.'[102] Contemporary humanity, in fact, feels *inferior* to their machines because, in 'their attempt to measure up to their machines', they realise that 'they are actually a "poor quality" raw material . . . Instead of actually being raw material, humans are "unblessed" morphologically because they are already pre-given.'[103]

It is worth noting that, in some ways, Jonas and Arendt's fear that the human being was being essentially diminished by its transformation into mere material is also present in the work of Heidegger, from the late 1930s and early 1940s. 'The power of machination – the eradication even of Godlessness, the anthropomorphizing of the human being into the animal, the exploitation of the earth, the calculation of the world – has passed over into a state of definitiveness,' he wrote.[104] By this time, of course, both Arendt and Jonas were well out of his sphere of influence; indeed, both would soon develop approaches which explicitly stood in opposition to Heideggerian philosophy and the nihilism they believed it invoked.

Jonas, as Lawrence Vogel explains, sought to extend Heideggerian categories in such a manner as to offer a more complete and thus existentially *and* ethically viable notion of being, namely to incorporate biological fact into the idea of being. 'Jonas charges that Heidegger's fundamental ontology does not fulfil its promise to delineate *transhistorical* structures of human existence,' Vogel explains. 'Instead, it bears testimony to the particular, historically fated situation of modern humanity: a situation defined by the "spiritual denudation of [the concept of nature] at the hand of physical science" since the Copernican revolution'.[105] While phenomenology had much to offer to the aspiring philosopher, Jonas explains, 'its restrictive emphasis on pure consciousness' is a subject of concern. 'What about the existence of our *body*?' he writes.[106] Heidegger never mentions the body, he notes; 'care' is not traced back to it:[107]

> German philosophy with its idealistic tradition was too high-minded to take this into account. Thus Heidegger too failed to bring the statement 'I am hungry' within the purview of philosophy . . . By ignoring the concrete basis of *ethics*, Heidegger's interpretation of inwardness denied itself an important means of access to this field; with this lack, ethics for him remained empty of real content. It was crucial for human beings to choose, but *what* choices they should make was not stated . . . Behind all of this, quite apart from Heidegger's particular case, lay an age-old one-sidedness from which philosophy has suffered: a certain disdain for nature, to which the mind or spirit felt itself superior. This was the heritage of the metaphysical *dualism* that has polarized Western thought since its origins in Platonism and Christianity.[108]

Arendt, on the other hand, found issue with Heidegger's failure to consider the political as a meaningful space. Being as *Dasein* is an essentially unpolitical creation in Heidegger's work. His failure to understand the essentially plural character of the human condition is one of Arendt's great critiques of Heideggerianism, as is her incorporation of that other Arendtian category of 'natality', whereby she centred human meaningfulness not around mortality, but rather the human capacity to create anew.

While Heidegger failed to incorporate the body and its needs into his work, he did express concern in his later writings on technology of the transformation of the natural world into the material of production, namely stored-up energy, and the inclusion of Being in this, as another kind of raw material. Other critics of technology characterise the deterioration of the human person in other ways. 'The individual transformed into a mass-produced being is left with only a numerical identity,' writes Anders. But now this has been lost, he argues, 'the numerical remainder is itself *"divided", the individuum has been transformed into a "divisum,"* broken into a multiplicity of functions. Manifestly, the destruction of man cannot go further; we cannot become any more inhumane.'[109] Our bodies, he writes, 'have become universal property'.[110] Declining respect for life has now been reflected back upon ourselves, Horkheimer argues, in the destruction of our inner life. This is 'the penalty man has to pay for having no respect for any life other than his own. The violence that is directed outward, and called technology, he is compelled to inflict on his own psyche.'[111] In *One-Dimensional Man*, Marcuse argues that 'this civilization transforms the object world into an extension of man's mind and body . . . Mass production and mass distribution claim the *entire* individual.'[112]

There is a belief consistently evident here that humans are, increasingly, being turned into things: mere bodies, mere life, raw material. The work of Arendt and Jonas offers a more developed analysis of what this means for politics and society in modernity: the implications for the 'human rights' of non-citizens and the manufacture of corpses that is uniquely enabled by the stripping away of humanity from humans, and, on the other side of the coin, the implications for those whose bodies are worked upon, 'healed' or improved by the technologies of creation. In both cases, the historical desacralisation of the body bears much responsibility, as Arendt and Jonas are both clear. But, for both, technologisation builds upon and extends this desacralisation into new realms: bodies are now increasingly the material upon which technology works, for the 'good' of the whole; we are increasingly modelled upon the form of machines, at the expense of the human.

NOTES

1. David Rorvik, *Brave New Baby* (London: New English Library, 1978), 1.
2. Michael Behrent, 'Foucault and Technology', *History and Technology*, 29:1 (2013).
3. Theodor W. Adorno and Max Horkheimer, *Dialectic of Enlightenment* [1944/47] (London: Verso, 1997), 232–3.
4. Ibid., 234.
5. Hans Jonas, 'Prologue', in *Mortality and Morality: A Search for Good after Auschwitz*, ed. Lawrence Vogel (Evanston: Northwestern University Press, 1996), 50.
6. Ibid.

7. Ibid., 51.
8. Dana Villa, *Arendt and Heidegger: The Fate of the Political* (Princeton: Princeton University Press, 1996), 192.
9. Günther Anders, *Die Antiquiertheit des Menschen*, vol. 2: *Über die Zerstörung des Lebens im Zeitalter der dritten industriellen Revolution* (Munich: Verlag C. H. Beck, 2018), 23.
10. Hannah Arendt, *The Origins of Totalitarianism* [1951] (Orlando: Harcourt, 1968), 297.
11. Ibid.
12. Ibid., 293.
13. Ibid., 294.
14. Ibid., 296.
15. Ibid., 297.
16. Ibid., 299.
17. Ibid.
18. Ibid.
19. Ibid., 298.
20. Ibid., 299.
21. Ibid., 302.
22. Hannah Arendt, *The Human Condition* (Chicago: University of Chicago Press, 1998), 2.
23. Ibid., 3.
24. Ibid., 263.
25. Ibid., 147.
26. Ibid., 153.
27. Ibid., 313–14.
28. Ibid., 314.
29. Ibid., 316.
30. Ibid., 318–19.
31. Ibid., 312.
32. Ibid., 313.
33. Ibid., 269.
34. Ibid.
35. Hannah Arendt, 'The Concept of History' [1961], in *Between Past and Future* (New York: Penguin, 2006), 59.
36. Ibid.
37. Hans Jonas, 'Biological Engineering – A Preview', in *Philosophical Essays* (New York and Dresden: Atropos Press, 1980), 144.
38. Hans Jonas, 'Seventeenth Century and After: The Meaning of Scientific and Technological Revolution', in *Philosophical Essays* (New York and Dresden: Atropos Press, 1980), 78.
39. Ibid., 81.
40. Ibid.
41. Hans Jonas, 'Technology and Responsibility: Reflections on the New Tasks of Ethics', in *Philosophical Essays* (New York and Dresden: Atropos Press, 1980), 15; 16.
42. Ibid., 17.

43. Ibid., 19.
44. Hans Jonas, 'Against the Stream: Comments on the Definition and Redefinition of Death', in *Philosophical Essays* (New York and Dresden: Atropos Press, 1980), 140–1.
45. Ibid., 141.
46. Ibid., 132fn.
47. Ibid.
48. Ibid.
49. Ibid.
50. Ibid.
51. Hans Jonas, 'Introduction', in *Philosophical Essays* (New York and Dresden: Atropos Press, 1980), xiii.
52. Ibid.
53. Ibid., xiv.
54. Hans Jonas, *The Phenomenon of Life* (New York: Harper & Row, 1966), 25.
55. Ibid., 55.
56. Ibid., 60.
57. Ibid.
58. Jonas, 'Introduction', in *Philosophical Essays* (New York and Dresden: Atropos Press, 1980), xvii.
59. Hans Jonas, 'Philosophical Reflections on Experimenting with Human Subjects', in *Philosophical Essays* (New York and Dresden: Atropos Press, 1980), 108.
60. Jonas, 'Technology and Responsibility', 14.
61. Hans Jonas, 'Contemporary Problems in Ethics from a Jewish Perspective', in *Philosophical Essays* (New York, Dresden: Atropos Press, 1980), 170.
62. Ibid.
63. Ibid., 171.
64. Ibid., 172.
65. Ibid., 173.
66. Ibid., 174.
67. Ibid., 176.
68. Giorgio Agamben, *Remnants of Auschwitz: The Witness and the Archive* (Cambridge, MA: MIT Press, 1998), 68.
69. Giorgio Agamben, *Where are We Now? The Epidemic as Politics* (London: Urtext, 2021), 29.
70. Ibid., 49–50.
71. Arendt, *The Human Condition*, 42.
72. Ibid., 43.
73. Ibid.
74. Ibid., 45.
75. Ibid.
76. Ibid., 46.
77. Ibid., 322.
78. Jonas, *The Phenomenon of Life*, 122.
79. Ibid., 126.
80. Ibid., 214.

81. Max Horkheimer, 'Threats to Freedom', in *Critique of Instrumental Reason* (London: Verso, 2012), 157.
82. Ibid., 152.
83. Marcuse, *One-Dimensional Man*, 12.
84. Ibid., 27.
85. Ibid., 33.
86. Ellul, *The Technological Society*, 382.
87. Ibid., 382–3.
88. Ibid., 383.
89. Ibid.
90. Ibid.
91. Ibid.
92. Ibid., 384
93. Anders, *Die Antiquiertheit des Menschen*, vol. 2, 116.
94. Ibid., 10.
95. Theodor W. Adorno and Max Horkheimer, *Dialectic of Enlightenment* [1944/47] (London: Verso, 1997), 131.
96. Arendt, *The Human Condition*, 312–13.
97. Hans Jonas, *Memoirs: Hans Jonas* (Waltham: University Press of New England, 2008), 203.
98. Lewis Mumford, *The Myth of the Machine: The Pentagon of Power* (New York: Harcourt Brace Jovanovich, 1970), 57.
99. Herbert Marcuse, 'Liberation from the Affluent Society' [1967], in *The Dialectics of Liberation*, ed. David Cooper (Harmondsworth and Baltimore: Penguin, 1968), 175.
100. Ibid., 176.
101. Jonas, 'Contemporary Problems in Ethics', 183.
102. Günther Anders, 'On Promethean Shame', in *Prometheanism: Technology, Digital Culture and Human Obsolescence*, ed. Christopher John Müller (London and New York: Rowman & Littlefield, 2016), 39.
103. Ibid., 51.
104. Martin Heidegger, *Ponderings XII–XV* [1939–41], trans. Richard Rojcewicz (Bloomington and Indianapolis: Indiana University Press, 2017), 41.
105. Lawrence Vogel, 'Introduction', in *Mortality and Morality: A Search for Good after Auschwitz*, ed. Lawrence Vogel (Evanston: Northwestern University Press, 1996), 8.
106. Hans Jonas, 'Prologue', in *Mortality and Morality: A Search for Good After Auschwitz*, ed. Lawrence Vogel (Evanston: Northwestern University Press, 1996), 43.
107. Ibid., 47.
108. Ibid.
109. Günther Anders, *Die Antiquiertheit des Menschen*, vol. 1: *Über die Seele im Zeitalter der zweiten industriellen Revolution* [1956] (Munich: Verlag C. H. Beck, 2010), 141.
110. Anders, *Die Antiquiertheit des Menschen*, vol. 2, 159.
111. Max Horkheimer, *Dawn and Decline* (New York: The Seabury Press, 1978), 161–2.
112. Marcuse, *One-Dimensional Man*, 9–10.

7

TECHNOLOGY AND WORLDLINESS: NATURE AND THE TECHNOLOGICAL ARTIFICE

INTRODUCTION

The critics of technology identified fundamental political problems in almost every form of technological activity in the contemporary world. Because of this pervasiveness, technology comes to form a new kind of world; it moulds the spaces in which we live: our environment. This chapter will outline the ways in which the critics of technology argue that technology is influencing worldliness, in constructing an increasingly all-dominating world of technological artifice, and the nature of that transformation from the pre-technological world. This also indicates the kind of future they believe we can anticipate.

The technological world comprises a unity or totality, embodying a new economics, politics, culture and so forth. But it also, very importantly for all the thinkers whose work has been analysed here, entails an extremely worrying relationship with 'nature', an idea that has been seen, in one way or another, in each of the previous chapters of this book. These philosophers have different ideas of what nature, in its authentic sense, means, whether that refers to human nature, the physical environment, or the relationship between the two. For some, the natural world offers a model for good living, or a refuge from the technological; for others, nature's force and power are something to be feared. Despite these differences in perspective, though, a common claim can be found among the critics of technology: that technology is divorcing humans from 'nature' as the basis or grounding of the human condition, and that this holds unpredictable and extraordinary ethical, political and existential risks. From the technology

186

critique, therefore, a common claim emerges about a changing natural environment and the role of human action within and upon that process. This chapter seeks to read this re-evaluation of nature in modernity as a political issue for these thinkers that demands and necessitates a political response.

The Concept of 'Nature'

The idea that modern technology is attacking 'nature', or that it has undertaken the conquest of nature, emerges again and again in the work of the critics of technology. Among all these theorists, a degree of consensus seems to emerge that the most foundationally transformative aspect of the contemporary technological world is the way it alters our relationship with nature, and this was by no means limited to the critics of technology. 'Anxiety about the triumph of the made over the grown was shared by a slew of twentieth-century figures across virtually every domain of thought and culture,' writes Benjamin Lazier, 'and it has hardly abated since.' The field of environmental history, he explains, 'is rooted in this anxiety'.[1] Yet the characterisation of nature offered by the critics of technology is not at all clear, even taking each the ideas of each theorist individually; the picture is even more confusing when they are brought together. Before moving on to their critique of the way that technology has transformed or is transforming nature and the relationship of humans to nature, this section will therefore outline the discordant ideas of 'nature' that can be found or are intimated in the work of these thinkers.

The theorists of technology hold different ideas about the character of nature: whether there is such a thing as authentic nature or whether all concepts of nature are constructed and historical; whether nature represents a model to follow or a danger to avoid; or whether nature represents greater risk or offers a stable and predictable foundation. Nature can also be seen in terms of (internal) human nature or (external) environment, although these two run into one another, as will be seen.

All the theorists accept that nature can be and has been understood in different ways in different eras. Despite appearances, nature is neither universal nor timeless. Their argument that technologised modernity imposes a new understanding of nature upon the contemporary world, and produces a new kind of relationship between humans and nature, necessitates this. They disagree, however, on whether an 'authentic' nature can be identified or whether 'nature' is always necessarily understood in relative and historically contextualised terms. Arendt, for instance, identifies certain key qualities within nature (as it relates to humans) in her insistence that boundaries be drawn between the political and the natural sphere: nature is inherently dangerous and extraordinarily powerful. Nature imposes certain demands upon humans: the necessity of labouring in order to live, the circularity and ceaselessness of that labour representing the immortality and relentlessness of nature itself.

Heidegger insists that nature can never be foundational to existence, but only a mode by which *Dasein* interprets being. The world as authentic being is ontological; nature is merely superimposed upon this. 'What is "natural" is *not* "natural" at all,' he writes, 'meaning self-evident for any given ever-existing man. The "natural" is always historical.'[2] Yet Heidegger's critique of the impositions of industrialisation upon the natural environment, not to mention the reverence with which he discusses the German pastoral does imply a preference for a particular idea of nature in a romantic mould. Nature appears as that from which we originate, the starting point, or a shelter. In Heidegger's 1935 'The Origin of the Work of Art', nature appears in the guise of earth, defined against world. The 'world', he writes, as 'the self-opening openness of the broad paths of simple and essential decisions in the destiny of a historical people', is contrasted with earth: 'that in which the arising of everything that arises is brought back – as, indeed, the very thing that it is – and sheltered. In the things that arise the earth presences as the protecting one.'[3] If world relates to historical existence, earth here describes nature as a protecting and sheltering foundation. The two are intimately related to one another.

Others suggest, through their writings, that there is or was an authentic nature in the sheer fact of the physical environments that could be found outside the city in the pre-modern era (Jonas), or before the ravages of industrialisation (Marcuse). Ellul, too, while maintaining that nature is a historically specific term, cites the significance of the 'natural environment' (which he notes is 'what is vulgarly called "nature": countryside, forests, mountains, oceans, etc.') for the determination of human nature and natural societies in a pre-technical world.[4] Yet these ideas remain somewhat ambiguous. One thinker who unambiguously holds that there is an authentic nature (to which we should aim to return) was Mumford. For Mumford, nature is conceived in terms of the balance realised within the organism, either individual or mass (as in ecosystems). This concept of nature can also be applied to society: a balanced ecosystem should be the model for human society, he argues.[5]

Against this, Adorno and Horkheimer offer perhaps the most relativistic analysis of nature of all the critics of technology. While they describe pre-modern ideas of nature in terms of myth, meaning nature was understood as something magical, embodied or represented in magical forms, contemporary ideas of nature are equally untrue, because in rationalising and objectifying nature we also misrepresent it. As we move from myth to enlightenment, they argue, we move from nature into 'mere objectivity', thus alienating ourselves from nature even as we increase our power over it.[6] As the 'magical illusion' fades, humans become trapped in the no less mythic idea of natural law, they argue.[7]

It is already evident that some of these theorists characterise nature as something to be feared, while others see it as a state or model to aspire to. Among those holding more positive, in some ways romantic, conceptualisations of nature are Mumford, Heidegger and Jonas. We should think of Mumford in the 'American tradition of worldly romanticism', argues Carl Mitcham.[8] In technological societies, writes Mumford, there is only one 'all-important mission in life: to conquer nature . . . to remove all natural barriers and human norms and to substitute artificial, fabricated equivalents for natural processes'.[9] A more romantic conception of nature is also reflected in Heidegger's critique of the influence of the hydroelectric plant upon the Rhine, transforming the river through the technological essence of power station, into nothing more than a 'water power supplier'.[10] The same is true of his observation of how 'the cultivation of the field has come under the grip of another kind of setting-in-order, which *sets* upon nature' through technology.[11] But it is nowhere more apparent than when he talks about his own experiences of nature. When Heidegger was offered a position at the prestigious University of Berlin, 'he refused, citing an "inner relationship of [his own] work to the Black Forest and its people"'.[12]

> Strictly speaking I myself never observe the landscape. I experience its hourly changes, day and night, in the great comings and goings of the seasons. The gravity of the mountains and the hardness of their primeval rock, the slow and deliberate growth of the fir-trees, the brilliant, simple splendor of the meadows in bloom, the rush of the mountain brook in the long autumn night, the stern simplicity of the flatlands covered with snow – all of this moves and blows through and penetrates everyday existence up there, and not in forced moments of 'aesthetic' immersion or artificial empathy, but only when one's existence stands in its work. It is the work alone that opens up space for the reality that is these mountains. The course of the work remains embedded in what happens in the region.[13]

Despite his claims that the concept of nature develops historically, he clearly intimates through his reflections on the value of nature to him personally, as well as what the natural world has *lost* through technology's exploitation, that there is some kind of intrinsic value in nature.

Jonas's work contains perhaps the most extensive and considered articulation of nature of all of the critics of technology presented here. He positioned his understanding of nature directly against that of Heidegger, and the perceived exclusion of the organic from Heidegger's philosophy (here, Jonas's attacks on Heidegger focus on the earlier ideas of *Being and Time*, and not Heidegger's later work on technology or thing-ness).[14] Metabolism, writes Jonas, is the foundation of human existence, and describes a particular relationship between humans and environment. We can move beyond the (Cartesian) 'indifference of the mere form-matter relation' through the concept of 'need', he argues.[15] Human entities have a 'need for constant self-renewal, and thus need for the matter required in that renewal, and thus need for "world".[16] World, in the sense offered here, can be considered as natural environment and its resources. This is, of course, true of all animals, metabolism is a quality of life itself, but in the human animal, which has become self-conscious, metabolism also contains 'the cardinal polarities' on which more complex moral, psychological and other human qualities evolve and depend.[17] Our interaction with the natural environment is therefore primary to understanding what it is to be human. For Jonas, too, the natural world, and the human relationship with the natural world, can provide a kind of model of action into and against a technological future. The Greek *polis*, he writes approvingly, 'has almost the naturalness of an organic fact' compared with the 'extreme artificiality of our technologically constituted, electronically integrated environment and corresponding habits'.[18]

Arendt, however, stood in absolute opposition to her friend Jonas, who believed that 'man's fundamental responsibility could be explained biologically, on the basis of the natural order', or to Mumford and his belief that 'organisms' – which he uses to refer equally to societies *and* individual humans – 'are nothing less than delicate devices for regulating energy and putting it at the service of life'.[19] For Arendt, politics – the most meaningful and human activity a person could undertake – was, as much as anything else, the construction of an artifice differentiating or separating the human world from the natural.[20] It is only the boundaries of politics which protected the human world from the everpresent threat posed by nature.[21] Nature is always threatening, always more powerful than humans, and to exclude its overwhelming force from dominating *all* human activities – to clear a space for freedom – requires a continual and concerted effort.

In Adorno and Horkheimer's work, nature also appears as something to be feared, as a force of extreme domination. Human beings exist, in their natural state, in subjection to nature.[22] Their story of the Enlightenment is that of humanity's efforts to reverse this situation through becoming dominant over nature, which includes humanity itself. To use nature *for* domination 'is the only aim'.[23] But through enlightenment, nature has not been tamed, but rather unleashed upon the human world. 'Civilisation is the victory of society over nature which changes everything into pure nature.'[24] In this sense, the two are in agreement with their philosophical adversary Arendt.

Others are more ambivalent. 'The natural does not have an eminent and normative value for me,' writes Ellul.[25] Yet, equally, the 'technical milieu is no more comprehensible . . . no more reassuring, no more meaningful than the "natural" milieu', and in the shift to the technological society, man 'feels the urgent need to get his bearings, to discover meaning and an origin, an authenticity in this inauthentic world', he argues.[26] While Marcuse adopts elements of Adorno and Horkheimer's characterisation of nature – describing the domination of technics as 'mastery of nature insofar as it is a hostile, violent, and destructive force; mastery of man to the extent that he is a part of that nature' – he also views nature in a more positive light, observing that the restoration of nature comprises part of the 'organic needs for the human organism'.[27]

There is great diversity and profound disagreement between these accounts of 'nature' in the pre-modern era. Some of these theorists view nature as essentially threatening, dominating, ceaseless and powerful; others understand nature as a protective, foundational space; some believe nature is, or should be, understood to be at the heart of human society and politics; others believe it is essential that nature be held at arm's length and defended against. The very extent to which nature can be defined is questioned, with some theorists appearing to offer some idea of an 'authentic' nature and others maintaining it can only be understood in particular historical contexts – with many offering *both* of these conflicting views, at different points in their work. There is no continuity or agreement in the conceptions of nature the critics of technology hold. If we look at the way in which they argue modern technology is transforming nature and our relationships with nature in the contemporary world, however, their arguments converge: the technological world is a corruption of the natural world.

The Technological World as 'Second Nature'

Technology, argue its critics, has created a new world. Lazier writes that, for Heidegger and Arendt:

> their vocabulary of earthliness recalls a category to which historians are accustomed: 'environment,' understood broadly as that which surrounds and conditions us. And their vocabulary of worldliness echoes in a second category: 'space of experience', which historians use to speak of urban cityscapes, changing skylines, and the new sorts of sensory and mental lives they afford or inflict.[28]

For Heidegger and Arendt, along with the other critics of technology, technology changes our relationship with the natural world. It creates new worldly – and sometimes unworldly – environments. 'Man no longer seeks to know the natural environment as such,' writes Ellul. It has been profoundly diminished.[29] Through technology we seek, rather, to escape the earth and our human condition, Arendt argues.[30] Modernity has filled the world with 'new organisms', Mumford claims, 'devised to represent the realities of physical science. Machines – and machines alone – completely met the requirements of the new scientific method and point of view.'[31] Nature is thus displaced by an artificial replica in the machine.

The creation of new political, social, cultural or civilisational 'worlds' is not unique, of course; the rise of technology does not mark an original transformation in this sense. But there is a way in which technologisation is seen to be an entirely unconventional kind of historical transformation, and this is perhaps best expressed in the way in which many of the critics of technology describe the technical world as a 'second nature'. Marcuse and Anders both use this specific term, as earlier used by Lukács, but the concept that technology is constituting a world that has transcended and replaced (or which attempts to replace) nature can be found more widely, in Mumford, Jonas, Arendt, Heidegger and, particularly extensively, in Ellul, *despite* the fact that their ideas of 'nature' are so diverse.

Lukács's idea of second nature is closely connected to his idea of reification that was earlier discussed. When the commodity form 'becomes the universal category of society as a whole', he writes, reification assumes 'decisive importance', and understanding of this is essential if men are to hope to 'liberate themselves from servitude to the "second nature" so created'.[32] This process by which the commodity form is universalised creates a new artificial world, as powerfully dominating as nature in its original form, and which replaces – or masks – nature in its original form. As Sheehan explains, this concept of second nature constitutes for Lukács 'a separate order like nature that integrates technological objects such as railroads and factory utensils as well as the relationships between human beings and the relationship between human beings, technological artifacts and nature'.[33]

Marcuse picks up this idea. The natural world, he argues, has been transformed into a technical world.[34] This world is all-encompassing: 'the technical world has crystallized into a "second nature," more hostile perhaps and more destructive than initial nature, pretechnical nature.'[35] Today, technology 'projects a historical totality – a "world"'.[36] Domination perpetuates itself 'not through but *as* technology'.[37] Anders reflects the same ideas when he writes that, in the nuclear era, 'we might speak of a "second nature", a term that had previously only been used metaphorically, but can now be used in a non-metaphorical sense, since there are now processes and pieces of nature that did not exist before they were created by us'.[38] Radio and television, he writes, 'produce a second world: the image of the world in which today's humanity believes it lives'.[39] Underlying these comments are an analysis of commodification that echo Lukács s ideas. Today, 'world' can only mean one thing, argues Anders, 'raw material . . . not only the embodiment of what something could be made from, but the embodiment of what we are obliged to make something out of', and the counterpart to this is that 'the existence of something from which nothing can be made must be denied'.[40] The existence of parts of nature – the Milky Way, for instance – from which no profit can be extracted becomes 'a metaphysical scandal, a material outrage'.[41] Like the Nazis' '"life unworthy of living", such things fall under the heading of "beings unworthy of existence"', he writes.[42]

Comparable ideas can be found in the work of other critics of technology. For Mumford, industrialisation, and later technologisation, is always an attempt to dominate nature. It thus divides man from his 'organic habitat'.[43] However, Thomas Sheehan has pointed out the way in which Mumford considers technology as not *just* a means of dominating nature but rather frames it as a '"counterfeit," to the world of nature', a 'process of resynthesizing nature', an idea Sheehan argues is analogous to the second nature concept.[44] Jonas refers to the 'second, determinate nature' of the self-imposed technological 'necessity'.[45] Lazier has written of how the planetary perspective enabled by the *Earthrise* images of the late 1960s, and beneath this by the development of space flight, worried thinkers such as Arendt and Heidegger, in part because 'what appears as the Whole Earth is in fact just another instance of the technological globe – and still worse, a technological globe that masks its fact'.[46] This is one example of the way in which a second, technological nature (or image of nature) was seen to be increasingly replacing pre-technological ideas of nature.

For Arendt, this is worrying because it further develops the twin processes of world-alienation (alienation from the political world) and earth-alienation (alienation from the physical world) that are shaping modernity. In the modern age, she writes, 'man, wherever he goes, encounters only himself. All the processes of the earth and the universe have revealed themselves either as man-made or as potentially man-made.'[47] This stands in contrast to nature as that which grows out of itself, as she elsewhere defines it.[48] 'These processes', she wrote:

> devoured . . . the solid objectivity of the given . . . in the situation of radical world-alienation, neither history nor nature is at all conceivable. This twofold loss of the world – the loss of nature and the loss of human artifice in the widest sense . . . has left behind it a society of men who, without a common world which would at once relate and separate them, either live in desperate lonely separation or are pressed together into a mass.[49]

Heidegger's description of the technological has clear affinities with the concept of second nature: in enframing, technology produces an idea of reality that masks authentic being. In his wartime notebooks, he writes that machination 'prescribes the corresponding basic notion of what has to count as "natural." Thus within the machination of beings, precisely the "natural" is subject to the arbitrariness of power and is the veiling of that arbitrariness.'[50] In the contemporary world, nature becomes merely an 'institution' through which 'the empowerment of machination plays out and is constantly secured'.[51] Both nature and history, he writes (in the late 1930s) have become 'set in place' because they are 'calculable', a radically different situation to the ancient Greek capacity to 'gather' and 'apprehend' experience 'within its openness'.[52]

Finally, Ellul's concept of the 'technological environment', which increasingly excludes the natural, offers us an image of second nature in other terminology. Technology is today 'the sole mediator now recognized', Ellul claims, such that the relationship between technology and man is now *nonmediated*.[53] These 'mediations are so generalized, extended, multiplied, that they have come to make up a new universe; we have witnessed the emergence of the "technological environment"'.[54] Because technological mediation is exclusive and all-encompassing:

> there are no other relationships between man and nature; the whole set of complex and fragile bonds that man has patiently fashioned – poetic, magic, mythical, symbolic bonds – vanishes. There is only the technological mediation, which imposes itself and becomes total. Technology then forms both a continuous screen and a generalized mode of involvement.[55]

Technology not only constitutes environment but human nature as well. 'Because technology infringes directly upon man's life,' Ellul writes, it comes to exact from him 'adaptations like those once demanded by the natural environment'.[56] It is also notable that Ellul's closest intellectual partner, Bernard Charbonneau, also adopts the language of second nature in his work. Through the development of science and technology, humankind has attempted to free itself from the forces of nature 'by developing a social supernature – or second nature', writes Charbonneau. Therefore, in modernity, Charbonneau argues, the 'contradiction between nature and human freedom grows unchecked', despite the fact that (he argues) the two are interwoven.[57]

The new technological world, the 'second nature' that technology forms, while not framed in identical terms by these thinkers, has a set of defining characteristics that can be found across the critiques of technology they offer, which reinforce the idea that technological modernity constitutes a kind of reified nature. The technological world is an *alienated* world; it is a *totalising* world; and it is a world of *false progress*.

Technology as an alienated world

The technological world, the critics of technology claim, is essentially an alienated world: a world that humans are fundamentally alienated from, to an extent that has never been seen before. Through technology, a 'false' world emerges, or rather, technology *is* a false world. While the writings on this theme by the Frankfurt School thinkers included here – Adorno, Horkheimer, Marcuse and Anders – are important to this, the association of a profound alienation with technological worldliness is not limited to them.

The human subject, write Adorno and Horkheimer, is increasingly excluded from the world. 'Men [now] expect that the world, which is without any issue, will be set on fire by a totality which they themselves are and over which they have no control.'[58] At the same time, 'interiority withers away', writes Horkheimer, replaced by other goals: 'technological expertise, presence of mind, pleasure in the mastery of machinery, the need to be part of and to agree with the majority or some group which is chosen as a model and whose regulations replace individual judgment.'[59] The subject in the technological world, thus, at the same time 'succumbs to an external determination and yet also becomes a subjectless thing, a sort of solid unity of hardened subjectivity, like the law that governs him', argues Adorno.[60] The individual, agrees Marcuse, 'has become entirely objective, the subject which is alienated is swallowed up by its alienated existence. There is only one dimension, and it is everywhere and in all forms.'[61]

As we have seen, this exclusion of the subject from the world is obscured by the construction of a false world, hiding the reality of such alienation. Media, such as TV, 'obscures the real alienation between people and between people and things', Adorno writes.[62] 'The totality of the processes of mediation, which amounts in reality to the principle of exchange, has produced a second, deceptive immediacy.'[63] Industrial culture, through technology, 'endlessly reduplicates the surface of reality'.[64] Via the authority of technological media, the image has come to be the principal category of our lives, claims Anders, and as such:

> we encounter the most important things as illusions and phantoms, i.e. in a trivialized, if not unrealized, version. Not as a 'world' (the world can be acquired simply by travelling and experiencing it), but as a consumer item delivered to your home . . . We are deprived of the ability to distinguish reality from appearance.[65]

This is a world of 'pre-digested' images: 'in every broadcast: there is no phantom sent that does not already have its "meaning," i.e. what we should think of it and feel, as an inherent element.' We fail to notice this, Anders writes, 'because the daily and hourly glut of phantoms that appear as a "world" prevents us from ever feeling hunger for interpretation, for our own interpretation; and because the more we are stuffed with this pre-digested world, the more profoundly do we forget this hunger'.[66] Technique in fact pretends to *be* world in order to disguise that it is in fact a method of domination, argues Anders. '*It is an instrument in the form of a pseudo-microcosmic model which in turn pretends to be the world itself.*'[67]

The better the technology, the less its mediation is visible. Don Ihde gives the example of a pair of spectacles: this technology is 'embodied' because it is 'maximally "transparent"', becoming integrated into one's experience of the body, and of the world. 'My glasses become part of the way I ordinarily experience my surroundings; they "withdraw" and are barely noticed,' he writes.[68] The result of such invisible incorporation of technology into the human environment, for Ellul, is that these environments become, instead of human-oriented worlds, instead 'pure reflection[s] of the technological system, of technological objects, of images and discourses which can only be technological images and discourses on technology'.[69] Anders writes of how 'banalization', in filling the world with 'images of seemingly familiar things', acts to hide the reality of alienation by presenting 'the world itself, including its most distant spatial and temporal regions, as a single gigantic home, as a universe of comfort'. It masks alienation, he argues, and in denying its reality 'clear[s] a space for its unconstrained action'.[70]

The image of the alienated world that these thinkers depict is one in which humans are not at home in the world, where they are merely functions, or raw material, or secondary to what 'really' matters: the technological world and its processes. In this world, technology appears as the authentic world: the relationships between people, between people and things, and between people and nature, or earth itself, are unimportant to the ends of technological practice and progress. It should be noted that, for all these theorists, alienation is always part of the human condition. Steven Vogel writes, for example, that, for many Frankfurt School thinkers, alienation 'is simply our ontological condition. Nature is *beyond* us and *other than* us . . . alienation cannot be overcome.'[71] Arendt's explanation of what she calls 'earth alienation' highlights the inexorably paradoxical nature of our relationship with our planet: earth alienation originates in the thoughts inspired by Galileo's telescope, when humans first started to think of their world from the perspective of space (as opposed to from a perspective of earthliness, or as grounded in the earth). At the beginning of the modern revolution in science and technology 'the old dichotomy between earth and sky was abolished and a unification of the universe effected, so that from then on nothing occurring in earthly nature was viewed as a mere earthly happening'. All events were considered to be 'universal', and subject to universally valid laws. Ever since, 'earth alienation became and has remained the hallmark of modern science'.[72] In better understanding earth as a whole – in understanding its position in relation to the rest of the universe, we become alienated from our (nonetheless still inevitable) condition as earthly creatures. Similarly, in our attempts to master nature, we now find ourselves at the mercy of unworldly (atomic) processes unleashed in the world. Certainly, there is no such thing as a world in which humans are not alienated from *something*, but the contemporary technological world is distinct in its degree of structural alienation: such a world is built for technology, in opposition to the worlds we build or have in the past attempted to build for ourselves.

Technology as a totalising world

The technological world collapses all spheres of life, culture, and society into one, it is argued. From this, technology's political character as totalitarian follows. The claim that different forms of action and modes of existence are increasingly indistinguishable, due to technology, appears in the work of the critics of technology in many different ways. As we have seen, it is claimed that the pressures of technology collapse the distinction between the public and the private spheres. We have experienced the '*de-privatization of the private sphere*,' argues Anders, while at the same time the 'public sphere has lost its definition . . . [and is now] seen as *only a continuation of the private sphere*.'[73] The private sphere has been 'invaded and whittled down by technological reality,' Marcuse observes. 'Production and mass distribution claim the *entire* individual.'[74] The division of the private and public, argues Arendt, so essential to premodern politics, has been erased by the rise of 'social' politics which merges the two spheres of life; they 'constantly flow into each other.'[75] In the technologies of production, or those of the media, to the behavioural sciences that accompanied the rise of modern technology, the individual has been engulfed in waves of standardisation and massification, while the various different modes of the public or political sphere have become increasingly indistinguishable through the same influences. We are rapidly moving towards the merging of all machines into a single apparatus, writes Anders, which 'assign to all existing things their function' and integrates all persons as its 'functionaries.' Even though this is 'not [yet] the case today . . . [and] the apparatuses are only on the way to this equation . . . they are already seen as "candidates", as parts of the "universal machine" that is currently being developed.'[76]

The universalism of the technological collapses the productive tension, dialectic, or difference between different parts and spheres of the world. Mediacy, or distance, is *essential* for the functioning of normal worldliness. The loss of distance through technology is highlighted in the work of several of these thinkers. The technologically produced unity of the earth represents its shrinking, and has been transformational to how humans understand and act in relation to the earth, writes Arendt.

> Men now live in an earth-wide continuous whole where even the notion of distance, still inherent in the most perfectly unbroken contiguity of parts, has yielded before the onslaught of speed. Speed has conquered space; and though this conquering process finds its limit at the unconquerable boundary of the simultaneous presence of one body at two different places, it has made distance meaningless, for no significant part of a human life – years, months, or even weeks – is any longer necessary to reach any point on the earth.[77]

Arendt famously characterised the political as a table, which gathers people together while at the same time distancing them from one another to maintain their plurality (rather than unity), an analogy to which Jonas also subscribed. 'To live together in the world means essentially that a world of things is between those who have it in common, as a table is located between those who sit around it.'[78] It is the simultaneous difference and distinction that the table maintains that makes it relevant here, and it is this characteristic that technology lacks.

'*All distances in time and space are shrinking*,' asserts Heidegger. Man 'now reaches overnight, by plane, places which formerly took weeks and months of travel. He now receives instant information, by radio . . . The peak of this abolition of every possibility of remoteness is reached by television.'[79] Yet 'in the way in which everything presences . . . despite all conquest of distances the nearness of things remains absent.'[80] The essay in which he writes these words offers a critique of the limited perspective of scientific understanding, which for Heidegger also means its embodiment in technologies. Scientific understanding does not bring us near to the things of the earth because it does not comprehend their multifaceted characters. The true worldliness of things (or the 'thingness' of things, as Heidegger writes) is not revealed by scientific understanding, rather, to believe in an instrumental, rationalist or scientific account of things is to be blinded to an authentic account of reality (hence why he argues that the 'nearness of things' remains absent). 'Men alone, as mortals, by dwelling attain to the world as world,' he argues. 'Only what conjoins itself out of world becomes a thing.'[81]

Anders objects to the 'attitude of the scientist', his objectivity that 'neutralize[s] the difference between what is nearby and what is remote' and argues that 'in a certain sense, every reader, radio listener, television consumer, and cinema watcher is today transformed into a vulgar double of the scientist: they too expect to have everything equally close and equally distant to them.'[82] This collapse of distance originates in technology: not just media but nuclear weapons mean that today 'every body is in deadly reach of everybody else.'[83] But when the outside world loses its distance, he argues, and penetrates the home, 'we no longer encounter it as "world."'[84] The loss of distance neutralises worldliness because it neutralises the distinctions that are necessary for meaning. Ellul, similarly, writes that 'tension between groups composing the entire society is a condition for life itself, or life susceptible to creation and adaptation in that society.'[85] No culture can exist, he argues, without such tension being maintained. 'To say this is no more than to affirm the reality of a certain dialectical movement in history,' a movement also, of course, considered essential by Marcuse, Adorno and Horkheimer.[86] For the critics of technology, the technological shift towards uniformity and totality, against distinction, difference, plurality and distance, comprised an inexorable drive in modern technology which threatened to smother dialectical movement altogether.

'Globe-talk' and the language of globalisation, Lazier argues, is an 'invention of the Earth-rise era', the era that saw the first images of the entire Earth from space.[87] The startling novelty of these images certainly provoked ideas of global unity, and connectivity, as well as of planetary isolation within space, but they also, as Lazier suggests, formed part of a 'history of competing global-isms', of which, we might add, the discourse on catastrophic technology was a major part.[88] The totalising nature of technology can thus be seen in the way that it has joined up – negatively unified – the globe and its peoples. 'Technology united the world,' writes Arendt. This means that 'for the first time in history all peoples on earth have a common present: no event of any importance can remain a marginal accident in the history of any other.'[89] Adorno likewise notes a 'kind of convergence towards a kind of universal standard at the level of technical rationality,' *particularly*, he argues, in 'countries which had previously been excluded from what Germans think of as the pull of universal history.'[90] The West 'has unified the world', declares Ellul. 'Western rational, efficient methods have taken over the world.'[91] He claims that one of the fundamental qualities of modern technology is technological universalism. 'It is easy to see that technique is constantly gaining ground, country by country and that its area of action is the whole world.'[92] This reverses the historical reality that previously, technologies belonged to individual civilisations. Instead, *today technique has taken over the whole of civilization.*'[93] The effect of this is to change everything. Political facts, for example, were formerly of two categories, those of local relevance – e.g. 'a local famine, a succession crisis in the local lord's family' – and political facts 'of general interest . . . not known to the entire population.'[94] But now, 'as a result of the global interconnectedness established by a network of communications systems, every economic or political fact concerns every man no matter where he may find himself.'[95] The technologised character of human action, echoes Jonas, 'changes the very nature of politics. For the boundary between "city" and "nature" has been obliterated, the city of men, once an enclave in the nonhuman world, spreads over the whole of terrestrial nature and usurps its place.'[96]

This represents, as Arendt writes, a negative solidarity, we must always remember, she argues, that 'technology, having provided the unity of the world, can just as easily destroy it.'[97] This reading of globalisation is in line with the critical reflection of many of these thinkers on Western influence in the international sphere: Arendt wrote extensively on the unprecedented brutality of imperialism in *The Origins of Totalitarianism*, Ellul highlighted the historical violence of American politics, in slavery, of course, but also the 'slow sanctimonious extermination of the Indians, [and] the system of occupying the land (*Faustrecht*)'.[98] Yet, it should be observed, this critique of technological colonialism, and the negative solidarity of technology places the critics of technology in tension with anticolonial thinkers such as Léopold Senghor or Frantz Fanon, who viewed technology and modernisation as a means of attaining liberation and political independence.[99]

The reading of the Western unification of the world through technology is also inflected with a particular critique of America's position in contemporary geopolitics. The philosopher George Grant, writing in 1968, whose work adopts a similar critique of technology to that outlined in this work, wrote that North Americans live 'in the most realised technological society which has yet been; one which is, moreover, the chief imperial centre from which technique is spread around the world.'[100] Divisions in politics, between left and right, 'are carried on within this fundamental faith [in technology].'[101] In the opposition of many of the critics of technology to (for example) American action in Vietnam; or more broadly, their repeated assertion that the politics of America is no different in substance to that of the Soviet Union, a strong sympathy with this perspective is indicated. Reinhart Koselleck argued that the Cold War 'is a result of European history. Europe's history has broadened, it has become world history and will run its course as that, having allowed the whole world to drift into a state of permanent crisis.' For Koselleck, technology plays a role in this: the technology of communications that has made 'all powers omnipresent, subjecting all to each and each to all;' or the technology of 'interplanetary space' the result of which could well be 'to blow up mankind . . . in a self-initiated process of self-destruction.'[102] The same kind of historical narrative is present in the histories of technology that its critics present: the development of technology, out of the European scientific and later industrial revolutions, enfolds the world, with its power centres moving both West and East into the US and USSR.

Technology as a world of false progress

Progress, wrote Heidegger, is only a principle of liberalism in appearance. 'In truth progress pertains to the essence of an age such as modernity which mistakes what is constantly new with what is genuinely true and real.'[103] Modern technological 'progress', wrote Adorno, from the slingshot to the atom bomb, is in truth 'completely inimical to the progress of freedom, to the advance of the autonomy of the human species', and such continued progression would be 'catastrophic'.[104] The orientation of modern progress is reflected also in Arendt's observation that, in the modern world of machines, 'even the most general end, the release of manpower, that was usually assigned to machines, is now thought to be a secondary and obsolete aim, inadequate to and limiting potential "startling increases in efficiency"'.[105] In a world shaped by machines, the concept of machine progress, and its prioritisation of efficiency over particular ends, is pervasive. The technological universe, in which we live today, prioritises a 'concern with effectiveness', notes Ellul.[106] 'The premise underlying this whole age', wrote Mumford, is 'the doctrine of progress'.[107] Technological change and human improvement became 'coupled together', in an ambition of man 'overcoming his physical limitations so as to impose his own machine-conditioned fantasies upon nature'.[108] The belief in progress – as described in this way – has masked the basic nihilism of technological motion, all agree. Thus, 'intensified progress seems to be bound up with intensified unfreedom', argued Marcuse.[109] 'The dual nature of progress', claims Adorno, 'always developed the potential of freedom simultaneously with the reality of oppression.'[110] While Adorno argues the status inherent in modern 'progress' originated in the exchange principle – that progress negates itself in its evolution – it 'intensifies by virtue of technology into the domination by repetition within the sphere of production. The life process itself ossifies in the expression of the ever-same.'[111]

This false notion of progress associated with technology has the effect of blinding us to the effects of our actions, these theorists argue. The mode of action of contemporary science and technology is illustrated, Arendt writes, by the scientists' remark that 'basic research is when I am doing what I don't know what I am doing'. When we begin, however, to make nature, to 'act into it' in Arendt's phrasing, or to unchain natural processes in the world, 'processes are started whose outcome is unpredictable, so that uncertainty rather than frailty becomes the decisive character of human affairs'.[112] We cannot 'undo' our actions in the world, she writes, and this is a timeless quality of human activity. Today, however, our 'incapacity to undo what has been done is matched by an almost equally complete incapacity to foretell the consequences of any deed or even to have reliable knowledge of its motives'.[113] Technology automates mental processes, Adorno and Horkheimer claim, 'converting them into blind cycles'.[114] In short, writes Ellul, the West is 'rushing nowhere at an ever increasing speed'.[115]

> We see the mistakes we have made, but we continue to make them with an apparently blind obstinacy. We know that there is an atomic threat and what it means, but like moles we go on building H-bombs and atomic energy plants . . . We are caught up in the madness and hubris of the dance of death: the important thing is the dance, the saturnalia, the bacchanalia, the lupercalia. We are no longer worried about what will emerge from it or about the void to which it points.[116]

Above all, technology's progressive facade is seen to represent a catastrophic failure to consider or comprehend the nature of the risk that contemporary technology poses to human existence.

THE TECHNOLOGICAL THREAT TO HUMANS AND THEIR ENVIRONMENT

It is already evident that, for the critics of technology, modern technology is a form of domination over nature as traditionally understood and over human freedom and even human nature. Domination, of course, is hardly unique to modernity, nor to technology. But the form of technological domination is original. As a totalising, all-consuming 'second nature', technology diminishes the status of both nature and humans, because both, the critics of technology agree, are now considered mere material for consumption, or 'raw material for products'.[117] Heidegger makes such a claim when he writes that the 'revealing' of modern technology 'challenges' nature, putting to it 'the unreasonable demand that it supply energy that can be extracted and stored as such'.[118] This, he argues, does not hold true for earlier technologies: 'the windmill does not unlock energy from the air currents in order to store it. In contrast, a tract of land is challenged into the putting out of coal and ore.'[119] In the present era, 'nature is no limit of technology . . . much more the fundamental piece of inventory of the technological standing reserve – and nothing else'.[120] Anders writes that, in modernity, 'the dignity that corresponds to the unique specimen is low'.[121] It is today a rule, he argues, that *"the unprofitable does not exist; or is not worthy of existence."* Our era shows with sufficient clarity that everything and anything . . . can become utterly worthless, condemned to become dregs: in this, people are no different from radium-contaminated nuclear waste.'[122] In *One-Dimensional Man*, Marcuse argues that 'only in the medium of technology, [do] man and nature become fungible objects of organization . . . Technology has become the great vehicle of *reification*.'[123] Nature has become 'mere objectivity', claim Adorno and Horkheimer.[124]

We are thus distanced from nature as it stood, independent of humanity, in the pre-technological era, while at the same time have invited the forces of nature into the human world, to utilise and reshape nature for human purposes. The too successful attempt to harness the power of nature is seen to have catastrophic effects on the world of human politics. Through technology, Jonas summarises, '*man* may have become more powerful, *men* very probably the opposite, enmeshed as they are in more dependencies than ever before . . . [Society's] compulsions, I fear, are almost as great as were those of unconquered nature.'[125]

But it is not simply, or even primarily, the extremes of technological domination that makes modern technology 'catastrophic' but rather the way in which it creates a state of extraordinarily heightened risk for humans and for their natural environments. The inherent risks in human action were limited, up to the contemporary era, Arendt writes, by the fact that it was confined to the human world. With the splitting of the atom, we 'started natural processes of our own . . . for the first time [we] have taken nature into the human world as such and obliterated the defensive boundaries between natural elements and the human artifice by which all previous civilizations were hedged in.'[126] Through the unchaining of nature, 'processes are started whose outcome is unpredictable, so that uncertainty rather than frailty becomes the decisive character of human affairs'.[127] Nuclear power represents the highest and most visible development of the risks of modern technology, but as the previous chapters of this book have shown, heightened risk to humans and their environments inheres in every realisation of technology.

Another aspect of the risk of technology – beyond the physical realities of particular technologies – is its pace of change. The acceleration of technology means that knowledge naturally acquired over time becomes irrelevant, and that it becomes impossible to acquire understanding of what we are doing at any given moment. The pace of chance renders historically acquired knowledge irrelevant. 'On the one hand we know more of the future than our premodern ancestors; on the other hand, we know far less,' Jonas explains. 'More, because our causal-analytic knowledge with its methodological application to the given is much greater; less, because we must deal with what is constitutionally a state of *change*, while those before us were dealing with what was . . . an abiding state.'[128] Technological risk is therefore inherent in the nature of modern technology. Just as significant, however, is the risk that is attached to the way that modern societies think about and relate to technology – arguably more so, because, while the genie of technology cannot be put back in the lamp, approaches to dealing with technology can be adapted.

It is the nature of modern technology, these thinkers agree, that the technological world reshapes people for that world, rather than improving the world for humanity. 'All-embracing technique is in fact the consciousness of the mechanized world,' says Ellul. It 'integrates everything . . . Man is not adapted to a world of steel; technique adapts him to it. It changes the arrangements of this blind world so that man can be a part of it without colliding with its rough edges.'[129] When an increasing number of problems are technological problems – because we live in an increasingly technological world – we come, incorrectly, to 'infer that all problems are technological', writes Ellul, and thus treat their solution as technological.[130]

Rationalisation, manipulability and calculability are of the essence of modern technology, but, writes Jonas, for the gain of calculability, there is a pay-off, and that was the emergence of 'neutrality', the absence of ends or final causes. As such, 'first Nature had been "neutralised" with respect to value, then man himself', he argues.

> Now we shiver in the nakedness of a nihilism in which near-omnipotence is paired with near-emptiness, greatest capacity with knowing least what for. With the apocalyptic pregnancy of our actions, that very knowledge which we lack has become more urgently needed than at any other stage in the adventure of mankind.[131]

Marcuse writes that technology 'considered as a historical process, is endowed with an internal meaning, a meaning of its own: it projects instrumentality as a means of freeing man from toil and anxiety.' Yet modern technique 'in the process of being developed as "pure" instrumentality, has disregarded this final cause, which no longer stands as the aim of technological development'.[132] Technology has lost its final (human) cause or end, and becomes 'pure' instrumentality because it is exclusively inward looking: oriented to the furthering of technological development for its own sake. 'Insofar as society tends to disregard the final cause of technology, technique in itself perpetuates misery, violence, and destruction.'[133] Heidegger characterises 'machination' as something that has compelled 'the whole of being into decisionlessness'.[134]

In the context of the absolute prioritisation of technological calculation, where 'pure instrumentality', or ideas of progress directed towards the development of technology (and not humans), dominate, it becomes impossible to calculate the risk that any technology, still less technology *as such*, poses to human life or society in any reasonable sense. This blinds human actors to technology's risks.

Closely related to this is the ideology of technological progress. 'Optimism, as confidence in man, in his powers and natural goodness, is the signature of modernity,' argues Jonas.[135] 'No leader of a revolution can afford to suspect the revenges which nature – human and environmental – may hold in store, in the immense complexity of things and the unfathomable abyss of the heart, for the planner of radical change.'[136] Marcuse also notes that we are prevented from understanding the reality of the contemporary destruction of natural resources. It is 'obscured and anaesthetised', he argues, 'by the fact that destruction itself is internally joined to production and productivity'.[137] It is abundantly clear that, in the narratives set out in this book, that technological progress is illusory, and that this illusion is part of its power. It is also part of the cause of the extraordinary risk that technology now poses.

This stands in absolute contradiction to the claims of technological optimists of this era. In 1964, IBM funded a major study into the effects of technological change on society and the influence of social change on the direction of technological development.[138] Emanuel Mesthene, the director of the Harvard University Program on Technology and Society, as it became, made the following case, in a summary of some of the program's findings, written in 1967. Our age is distinct in two ways, he wrote:

> first, we dispose, in absolute terms, of a staggering amount of physical power; second, and most important, we are beginning to think and act in conscious realization of that fact. We are therefore the first age who can aspire to be free of the tyranny of physical nature that has plagued man since his beginnings.[139]

Our technological capacity brings about new powers, new possibilities and more choices, he argues. 'With more choices, we have more opportunities. With more opportunities, we can have more freedom, and with more freedom we can be more human.'[140] It is precisely this faith in technological progress that the critics of technology speak against.

Technology's risk springs from the combination of its physical power, the rapidity of change and the corresponding irrelevance of historical knowledge and understanding, the centring of technology and its 'pure objectivity' at the expense of the human (with all that entails), and the illusion of technological optimism – that all things will be for the best. The potential for catastrophic risk has heightened exponentially while at precisely the same moment our capacity to understand, calculate and mitigate against those risks has collapsed. The 'second nature' that technology projects upon the world is a space predicated upon its *claim* to rationality, calculability and predictable order. Yet, the critics of technology agree, in attempting to subjugate nature, we unleash its potential for chaos and catastrophe. We have collapsed, in almost every imaginable sense, the boundaries, limits or balances that have traditionally existed in the relationship between humans and nature and which have ordered that relationship. And, the critics argue, we are, largely, blind to this fact, instead labelling it 'progress'.

The way that these thinkers use ideas of worldliness, and the concerns they have about the influence of technology on the world humans depend upon, resonates strongly with the modern concept of environment. Technology challenges the human world in (often) highly destructive ways, but at the same time it challenges the earth itself, and even the character and qualities of nature. Technology is depicted as a force that has formed its own world, and which even attempts to manipulate nature – successfully, to some degree, in that technology is increasingly breaking through the (historical) limits of nature in different ways. But the critics of technology all suggest, albeit with reference to different ideas of 'nature', that this attempt to manipulate nature through technology or by technology is likely to be catastrophic for both the social and political worlds of humanity and the earthly environment in which they live. This is because of the unnatural power that technology has brought into the world, realised in different ways as domination of human over human, but also because of the extraordinary and unprecedented degree of risk that technological world-making entails. The destruction of nature, human and environment, is inherent in technology as it is expressed here. 'Technology can become an environment only if the old environment stops being one,' writes Ellul.

> But that implies destructuring it as an environment and exploiting it to such an extreme that nothing is left of it. In other words: The well-known 'depletion' of natural resources . . . results not only from abuse by the technologies, but also from the very establishment of technology as man's new milieu.[141]

Notes

1. Benjamin Lazier, 'Earthrise; or, The Globalization of the World Picture', *The American Historical Review* 116:3 (2011), 603–4.
2. Martin Heidegger, *What is a Thing?* [1935], trans. W. B. Barton Jr. and Vera Deutsch (Chicago: Henry Regnery Company, 1967), 39.

3. Martin Heidegger, 'The Origin of the Work of Art', in *Off the Beaten Track*, trans. Julian Young and Kenneth Haynes (Cambridge: Cambridge University Press, 2002), 21.
4. Jacques Ellul, *The Technological System*, trans. Joachim Neugroschel (New York: Continuum, 1980), 38–40.
5. Lewis Mumford, *The Myth of the Machine: The Pentagon of Power* (New York: Harcourt Brace Jovanovich, 1970), 395–6.
6. Theodor W. Adorno and Max Horkheimer, *Dialectic of Enlightenment* [1944/47] (London: Verso, 1997), 9.
7. Ibid., 12.
8. Carl Mitcham, *Thinking through Technology: The Path between Engineering and Philosophy* (Chicago: University of Chicago Press, 1994), 40.
9. Mumford, *The Myth of the Machine: The Pentagon of Power*, 172–3.
10. Martin Heidegger, 'The Question Concerning Technology', in *The Question Concerning Technology and Other Essays*, trans. William Lovitt (New York: Harper & Row, 1977), 16.
11. Ibid., 14–15.
12. Casey Rentmeester, *Heidegger and the Environment* (Washington, DC: Rowman & Littlefield, 2015), 78.
13. Martin Heidegger, cited in Rentmeester, *Heidegger and the Environment*, 78.
14. For example, the language Jonas uses when referring to Heidegger's concept of worldhood and environment is quite clearly drawn from Heidegger's work of the 1920s. Thus, in 1977, Jonas wrote (quoting Arendt): '"To live together in the world means essentially that a world of things is between those who have it in common, as a table is located between those who sit around it." Please note how superior this strikingly simple observation is to what Heidegger had to say about the "worldhood" of the world. That, too, is constituted of tables, hammers, ploughs, but these are subsumed under the category of "implement" or "equipment", and the worldhood of the world is essentially its being a referral context. Neither the artificiality nor the durability nor the public commonality of this pragmatic web is stressed by Heidegger. For Hannah Arendt, the man-made world of things is more than a utility: it is (without any mystique of the "thing") the true human habitat, erected to last and to enclose, interposed not only between man and nature but also between man and man: in the latter case separating and uniting at the same time, as one meets over a table.' Hans Jonas, 'Acting, Knowing, Thinking: Gleanings from Hannah Arendt's Philosophical Work', *Social Research* 44:1 (1977), 32–3.
15. Hans Jonas, 'Biological Foundations of Individuality', in *Philosophical Essays* (New York and Dresden: Atropos Press, 2010), 196.
16. Ibid.
17. Ibid., 198.
18. Hans Jonas, 'Seventeenth Century and After: The Meaning of Scientific and Technological Revolution', in *Philosophical Essays* (New York and Dresden: Atropos Press, 2010), 80.
19. Lewis Mumford, *The Myth of the Machine: Technics and Human Development* (New York: Harcourt Brace Jovanovich, 1967), 344–51.

20. Hans Jonas, *Memoirs: Hans Jonas* (Waltham: University Press of New England, 2008), 203.
21. Hannah Arendt, *The Human Condition* (Chicago: University of Chicago Press, 1998), 125–6.
22. Adorno and Horkheimer, *Dialectic of Enlightenment*, 42.
23. Ibid., 4.
24. Ibid., 186.
25. Ellul, *The Technological System*, 47.
26. Jacques Ellul, *The New Demons*, trans. C. Edward Hopkin (New York: The Seabury Press, 1975), 66–7.
27. Herbert Marcuse, 'World without a *Logos*', *Bulletin of the Atomic Scientists* 20 (January 1964), 25; Herbert Marcuse, 'Liberation from the Affluent Society' [1967], in *The Dialectics of Liberation*, ed. David Cooper (Harmondsworth and Baltimore: Penguin, 1968), 186.
28. Lazier, 'Earthrise', 627.
29. Ellul, *The Technological System*, 39–40.
30. Arendt, *The Human Condition*, 2.
31. Lewis Mumford, *Technics and Civilization* (San Diego, New York and London: Harcourt Brace Jovanovich, 1963), 51.
32. Georg Lukács, 'Reification and the Consciousness of the Proletariat', in *History and Class Consciousness* (Cambridge, MA: The MIT Press, 1971), 67.
33. Stephen Sheehan, 'The Nature of Technology: Changing Concepts of Technology in the Early Twentieth-Century', *Icon* 11 (2005), 2.
34. Marcuse, 'World without a Logos', 25.
35. Ibid., 25.
36. Herbert Marcuse, *One-Dimensional Man* (Boston, MA: Beacon, 1964), 154.
37. Ibid., 158.
38. Günther Anders, *Die Antiquiertheit des Menschen*, vol. 2: *Über die Zerstörung des Lebens im Zeitalter der dritten industriellen Revolution* [1980] (Munich: Verlag C. H. Beck, 2018), 22.
39. Ibid., 60.
40. Ibid.
41. Günther Anders, *Die Antiquiertheit des Menschen*, vol. 1: *Über die Seele im Zeitalter der zweiten industriellen Revolution* [1956] (Munich: Verlag C. H. Beck, 2010), 184.
42. Anders, *Die Antiquiertheit des Menschen*, vol. 2, 35–6.
43. Mumford, *The Myth of the Machine: Technics and Human Development*, 3.
44. Sheehan, 'The Nature of Technology', 13.
45. Ibid., 49.
46. Lazier, 'Earthrise', 614.
47. Hannah Arendt, 'The Concept of History' [1961], in *Between Past and Future* (New York: Penguin, 2006), 90.
48. Arendt, *The Human Condition*, 150.
49. Ibid.
50. Martin Heidegger, *Ponderings XII–XV* [1939–41], trans. Richard Rojcewicz (Bloomington and Indianapolis: Indiana University Press, 2017), 6.

51. Ibid., 35.
52. Martin Heidegger, 'The Age of the World Picture', in *The Question Concerning Technology and Other Essays*, trans. William Lovitt (New York: Harper & Row, 1977), 127; 131.
53. Ellul, *The Technological System*, 34–5; 37–8.
54. Ibid., 38.
55. Ibid., 35.
56. Ibid., 39.
57. Bernard Charbonneau, *The Green Light: A Self-Critique of the Ecological Movement* [1980] (London: Bloomsbury Academic, 2018), 55; 60.
58. Adorno and Horkheimer, *Dialectic of Enlightenment*, 29.
59. Max Horkheimer, 'The Concept of Man', in *Critique of Instrumental Reason* (London: Verso, 2012), 12.
60. Theodor Adorno, *Against Epistemology: A Metacritique; Studies in Husserl and the Phenomenological Antinomies* [1956] (Cambridge, MA: MIT Press, 1986), 62–3.
61. Marcuse, *On- Dimensional Man*, 11.
62. Theodor Adorno, 'Prologue to Television', in *Critical Models: Interventions and Catchwords*, trans. H. W. Pickford (New York: Columbia University Press, 1998), 49–50.
63. Theodor Adorno, 'Late Capitalism or Industrial Society?,' in *Can One Live after Auschwitz?: A Philosophical Reader* (Stanford: Stanford University Press, 2003), 124.
64. Max Horkheimer, *Eclipse of Reason* (London: Continuum, 1947), 98.
65. Anders, *Die Antiquiertheit des Menschen*, vol. 2, 279.
66. Anders, *Die Antiquiertheit des Menschen*, vol. 1, 198.
67. Ibid., 164.
68. Don Ihde, 'A Phenomenology of Technics', in *Philosophy of Technology: The Technological Condition – An Anthology*, ed. Robert C. Scharff and Val Dusek (Oxford: Blackwell, 2003), 508.
69. Ellul, *The Technological System*, 37–8.
70. Anders, *Die Antiquiertheit des Menschen*, vol. 1, 124–5.
71. Steven Vogel, 'On Nature and Alienation', in *Critical Ecologies: The Frankfurt School and Contemporary Environmental Crises*, ed. Andrew Biro (Toronto: University of Toronto Press, 2011), 190.
72. Arendt, *The Human Condition*, 264.
73. Anders, *Die Antiquiertheit des Menschen*, vol. 2, 95.
74. Marcuse, *One-Dimensional Man*, 10.
75. Arendt, *The Human Condition*, 33.
76. Anders, *Die Antiquiertheit des Menschen*, vol. 2, 123.
77. Arendt, *The Human Condition*, 250.
78. Jonas, 'Acting, Knowing, Thinking', 32–3.
79. Martin Heidegger, 'The Thing', in *Poetry, Language, Thought*, trans. Albert Hofstadter (Harper & Row: New York, 1971), 163.
80. Ibid., 164.

81. Ibid., 180.
82. Anders, *Die Antiquiertheit des Menschen*, vol. 1, 123.
83. Günther Anders, 'Theses for the Atomic Age', *The Massachusetts Review* 3:3 (1962), 495.
84. Anders, *Die Antiquiertheit des Menschen*, vol. 2, 232.
85. Jacques Ellul, *The Political Illusion*, trans. Konrad Kellen (New York: Knopf, 1967), 217; Jacques Ellul, *Propaganda: The Formation of Men's Attitudes*, trans. Konrad Kellen and Jean Lerner (New York: Knopf, 1965), 217.
86. Ibid.
87. Lazier, 'Earthrise', 608.
88. Ibid.
89. Hannah Arendt, 'Karl Jaspers' [1958], in *Men in Dark Times* (London: Cape, 1970), 82–3.
90. Theodor Adorno, 'Universal and Particular', in *History and Freedom: Lectures 1964–5* (Cambridge: Polity Press, 2008), 14.
91. Jacques Ellul, *The Betrayal of the West* (New York: Seabury, 1978), 123.
92. Ellul, *The Technological Society*, 116.
93. Ibid., 128.
94. Ellul, *The Political Illusion*, 97.
95. Ibid., 97–8.
96. Hans Jonas, *The Imperative of Responsibility* (Chicago: University of Chicago Press, 1984), 10.
97. Arendt, 'Karl Jaspers', 83.
98. Jacques Ellul, *Violence: Reflections from a Christian Perspective*, trans. Cecelia Gaul Kings (New York: Seabury, 1969), 53.
99. Roxanna Nydia Curto, *Inter-tech(s): Colonialism and the Question of Technology in Francophone Literature* (Charlottesville: University of Virginia Press, 2016), 3–4.
100. George Grant, 'In Defence of North America' [1968], in *Technology and Empire* (Toronto: House of Anansi Press, 1969), 40.
101. Ibid., 30–1.
102. Reinhart Koselleck, *Crisis and Critique: Enlightenment and the Modern Pathogenesis of Modern Society* [1959] (Cambridge, MA: The MIT Press, 1988), 5.
103. Heidegger, *Ponderings XII–XV*, 215.
104. Adorno, 'Universal and Particular', 12.
105. Arendt, *The Human Condition*, 152.
106. Ellul, *Propaganda*, x.
107. Lewis Mumford, *My Works and Days: A Personal Chronicle* (New York and London: Harcourt Brace Jovanovich, 1979), 6.
108. Ibid.
109. Herbert Marcuse, *Eros and Civilization: A Philosophical Inquiry into Freud* (London: Ark, 1987), 4–5.
110. Theodor Adorno, *Minima Moralia: Reflections from Damaged Life* [1951] (London: NLRB, 1974), 146–7.
111. Adorno, *Critical Models*, 159–60.

112. Arendt, *The Human Condition*, 232.
113. Ibid., 233.
114. Adorno and Horkheimer, *Dialectic of Enlightenment*, 181.
115. Ellul, *The Betrayal of the West*, 197.
116. Ibid.
117. Anders, *Die Antiquiertheit des Menschen*, vol. 1, 184.
118. Heidegger, 'The Question Concerning Technology', 14.
119. Ibid., 15.
120. Martin Heidegger, 'Insight into That Which Is' [1949], in *The Bremen and Freiburg Lectures*, trans. Andrew J. Mitchell (Bloomington and Indianapolis: Indiana University Press, 2012), 41.
121. Anders, *Die Antiquiertheit des Menschen*, vol. 1, 183.
122. Ibid., 184.
123. Marcuse, *One-Dimensional Man*, 168.
124. Adorno and Horkheimer, *Dialectic of Enlightenment*, 5–6.
125. Hans Jonas, 'Toward a Philosophy of Technology', *The Hastings Centre Report* 9:1 (1979), 201.
126. Arendt, 'The Concept of History', 60.
127. Ibid.
128. Jonas, *The Imperative of Responsibility*, 119.
129. Ellul, *The Technological Society*, 6.
130. Ellul, *The Technological System*, 49.
131. Hans Jonas, 'Technology and Responsibility: Reflections on the New Tasks of Ethics', in *Philosophical Essays* (New York and Dresden: Atropos Press, 1980), 19.
132. Marcuse, 'World without a Logos', 25.
133. Ibid.
134. Heidegger, *Ponderings XII–XV*, 41–3.
135. Jonas, 'Seventeenth Century and After', 75.
136. Ibid.
137. Herbert Marcuse, 'Ecology and the Critique of Modern Society', *Capitalism, Nature, Socialism* 3:3 (1979), 33–4.
138. Albert H. Teich, 'Introduction', in *Technology and Man's Future*, ed. Albert H. Teich (New York: St. Martin's Press, 1981), 4.
139. Emanuel Mesthene, 'Technology and Wisdom' [1967], in *Philosophy of Technology: The Technological Condition – An Anthology*, ed. Robert C. Scharff and Val Dusek (Oxford: Blackwell, 2003), 109.
140. Ibid., 111.
141. Ellul, *The Technological System*, 44.

CONCLUSION: THE LASTING INFLUENCE OF 'CATASTROPHIC TECHNOLOGY'?

INTRODUCTION

As Duncan Bell has written, 'the scope of the liberal tradition was massively expanded during the middle decades of the century, chiefly in the United States, such that it came to be seen by many as the constitutive ideology of the West.' This shift in understandings of liberalism, Bell observes, 'was produced by a conjunction of the ideological wars fought against "totalitarianism" and assorted developments in the social sciences'.[1] In the same period, faith in the advancement similarly promised by science and technology was also thriving. Isabelle Stengers argues that, in the mid-twentieth century, when 'the grand perspectives on techno-scientific innovation were synonymous with progress, it would have been quasi-inconceivable not to turn with confidence to the scientists and technologists'.[2] The concept of catastrophic technology stands apart from and in some ways opposes these undoubtedly more widely held positions. And although the critique of technology was a minority view, neither can it be said to be merely on the periphery of political thinking, given the significance of the political theorists who made the critique, and the extent to which their work was read and drawn upon. Many other thinkers were influenced by or engaged with this critique, beyond those whose work has been referenced in in this book, some of whom have already been highlighted in the previous chapters.

Of the post-war expansion of liberalism, Bell writes that 'today we both inherit and inhabit it'.[3] The same can be said of the connection between technology and progress. Despite an abundance of contemporary criticism of liberalism (and 'neo-liberalism'), it is true that liberal ideals still shape political thinking to a very significant extent. Similarly, despite, for instance, the touted 'end of the expert' or other similar rejections of scientific and technological expertise, we still largely place our faith in the possibilities of science and technology for medical advancement, the mitigation of environmental damage, or economic growth, for instance. Against these still-prevalent paradigms, the concept of catastrophic technology has not gained ground and as a historical narrative has been rather overlooked. Its significance for particular thinkers has been understood (in some cases) but, generally, not the connection between them and its breadth as a countervailing theoretical position in political thought.[4] Yet against dominant liberal-progressive ideologies it offers a provocative contrasting perspective on the Cold War world, and perhaps on our own.

While the narrative of catastrophic technology as a historical theme has been overlooked, there are ways in which it has influenced certain strains of political thinking since the Cold War era and in the contemporary world. This concluding chapter will outline some of the most important ways the critique lives on, through (perhaps most significantly) the environmental movement, as well as contemporary political theory, based on some of the ideas explored here. There was a convergence, in the 1960s and 1970s, between the ideas of early environmentalists and the critics of technology that was significant for both groups. Given the significance of the environment for the critics of technology – both in an abstract sense, but also in the modern usage of the term as the physical earth, sea and sky which surrounds us and on which we depend – it is plausible to read the critique of technology as an early 'Anthropocenic' form of understanding. Finally, this chapter will end with some ideas on how we might approach the idea of catastrophic technology from a contemporary perspective, the limitations of the argument, and why, despite these limitations, it might still be useful to reflect upon the claims of the critics of technology in the twenty-first century.

CATASTROPHIC TECHNOLOGY AND THE ANTHROPOCENE

In Christophe Bonneuil and Jean-Baptiste Fressoz's 2013 *The Shock of the Anthropocene*, they make the argument that the concept of the 'Anthropocene', understood as a new era in which human action is catastrophically influencing the global environment, has a much longer history than has been recognised by widespread narratives about the Anthropocene. 'The grand narrative of the Anthropocene is . . . the story of an awakening', they write, of 'a long moment of unawareness, from 1750 to the late twentieth century, followed by a sudden arousal'.[5] We are therefore, so the narrative goes, the first generation to understand the impact of our activities on the earth system. Yet, Bonneuil and Fressoz write, 'the problem with these grand narratives of awakening, revelation or arousal of consciousness is that they are historically wrong'.[6] They highlight, for example, the critique of overproduction and consumptionism that took place from the 1950s to the 1970s, including 'Marcuse, Baudrillard, Guy Debord, Marshall Sahlins, Daniel Bell, Christopher Lasch, the *Limits to Growth*, Anne and Paul Ehrlich, David Riesman, John Kenneth Galbraith, Vance Packard, etc.', a cultural critique bound up with environmental critique.[7] 'The problem with the narrative of ecological awakening,' Bonneuil and Fressoz argue, 'according to which our generation is the first to recognize environmental disturbance and question industrial modernity, is that by obliterating the reflexivity of past societies it depoliticizes the long history of the Anthropocene.'[8]

The critique of technology, and particularly the conceptualisation of technological worldliness that comes from that critique, is, considered in this vein, one example of the longer historical awareness of the impact of human action – particularly via technology – on the environment. The critics of technology offer a political critique of a world, and a globally interdependent environment, which they understand to be constructed by human beings – materially and ideologically. The driving force behind the reshaping of the world is technology, they argue, and this force, as we have seen, is catastrophic. Thus, in the critique of technology, a significant critique in the history of political thinking of this era, albeit one that has received far less attention than one might expect, given the intellectual significance of its protagonists, we might read a particular kind of politically informed understanding of the 'Anthropocene': a world in which our environment is fundamentally artificial, rather than natural. This idea is important because of its significant role in the work of so many thinkers who have influenced, and continue to influence, our thinking about the political world. This includes those whose work has been discussed in this book, as well as many others. Among these others are many theorists of the environment. The critics of technology are connected to – they influence and are influenced by – more overtly 'environmental' works that are written during the later Cold War, particularly from the late 1960s and the 1970s onwards. The Anthropocenic character of the critique of technology is not coincidental but closely connected to the kind of environmentalism that led in a direct path towards the popularisation of the Anthropocene concept in the early twenty-first century.

The technology critique intersects with the rise of environmental thinking and environmental movements that took place in the 1960s and 1970s, represented by writers such as Rachel Carson, Barry Commoner and the authors of the *Limits to Growth* report, among others. Others have identified the relevance of some of these thinkers' ideas to environmental themes (including Bonneuil and Fressoz), such as Arendt on the concept of earthliness, Marcuse on the liberatory possibilities of nature, and, perhaps most importantly, Heidegger on the way in which nature is now considered to be nothing more than 'raw material' for human consumption. In fact, Heidegger's work, in retrospect, has come to be seen as highly relevant to contemporary discussions of environmentalism, as the rapidly increasing literature on the theme shows.[9] This chapter thus suggests that as a cohesive discourse, the critique of technology, and the thinkers and philosophers involved in that critique, offer one particularly important, and highly developed example of the way in which human action influences the environment, and a school of thought which can be seen to have direct influence on early environmental philosophy and political movements. Ecologists and environmentalists coming from scientific backgrounds found in the technology critique a political-social discourse that chimed with their own concerns about the interconnectedness of man and nature, and the influence of nuclear power. In turn, the critics of technology found in the ecologists' work an exemplification of the extensive risks that threatened humankind via technology, and in this new awareness also some possibility of hope for the future.

On this latter point, it is surprising, even paradoxical, that the critics of technology, who were so fiercely critical of modern rationalised and technologised science, found in the science of ecology and its practitioners and followers so much potential. Murray Bookchin wrote in 1965 that 'the critical edge of ecology is due not so much to the power of human reason . . . but to a still higher power, the sovereignty of nature'. We may still attempt to manipulate man and nature, he argues, but 'ecology clearly shows that the *totality* of the natural world – nature viewed in all its aspects, cycles and interrelationships – cancels out all human pretensions to mastery over the planet'.[10] These words, it seems, can be applied to the thinking of the theorists of catastrophic technology. Ecology, and its discovery of the unpredictability, uncontrollability and extent of the damage being caused to environment by humans, has the power to shock modern society into waking up from the dream of technological progress. And if contemporary science was seen as essentially technological by the critics of technology – driven by a rationale not only to understand but to conquer and manipulate nature – ecology counters this in its observations of the limits of human action over or in the face of nature. It is absent of the 'hubris of domination', as Marcuse framed the scientific-technological attitude.[11]

The description of the technological 'world' offered by the critics of technology, and their analysis of the influence of human action – via technology – on 'nature', makes it quite clear that, to return to Bonneuil and Fressoz's point, the recognition of environmental disturbance, and the radical questioning of industrial modernity, is certainly not original to the twenty-first century.[12] The concept of the Anthropocene, not using that particular term, but rather in the conceptualisation of worldliness and its transformation through technology, is a central concern in the work of each one of these thinkers. Their work also highlights the directly political nature of this transformation of world: both in the sense that these critiques are embedded within a particular political context of the post-Second World War/Cold War world, but also in the fundamental political problems that are claimed to inhere in this technological reshaping of the world.

CATASTROPHIC TECHNOLOGY AND ENVIRONMENTALISM

The influence of environmentalism on the critics of technology

As well as the critical technology discourse representing a form of Anthropocenic thinking preceding the coining of that phrase by some decades, there is also a direct intellectual connection between the critics of technology and the contemporary environmental and ecological movements that emerged (or became popularised) in the 1960s and 1970s. Barbara Ward and Rene Dubos wrote in their 1972 *Only One Earth* that:

> as we enter the last decades of the twentieth century, there is a growing sense that something fundamental and possibly irrevocable is happening to man's relations with both his worlds [i.e. natural and social] . . . Men seem, on a planetary scale, to be substituting the controlled for the uncontrolled, the fabricated for the unworked, the planned for the random. And they are doing so with a speed and depth of intervention unknown in any previous age of human history.[13]

These words echo the warnings of the critics of technology; the two critiques begin to run in parallel, and they also engage with one another.

Part of the affinity of these two discourses is the shared context in which they arose, in the era of the development and expansion of nuclear technologies. Both the critics of technology and environmentalists were preoccupied with the risks that nuclear technology introduced into the world. In the case of environmental writers, their fears struck a particular chord with the wider public, leading to the broader impact and influence of their work. Just as for the critics of technology, it was not primarily the threat of nuclear war that concerned the environmentalists, but rather the unknown nature of the consequences of nuclear energy: the scope and patterns of radioactive fallout after tests or nuclear accidents, or the effects of such radiation on humans and animals, and the movement of radiation through ecosystems or food chains. Fears about the unpredictability of the risks surrounding these novel technologies were at the heart of environmentalism in the early days, just as they were at the heart of the technology critique.

'What is happening', wrote Marcuse in 1972, 'is the discovery (or rather, rediscovery) of nature as an ally in the struggle against the exploitative societies in which the violation of nature aggravates the violation of man.'[14] Furthermore, he argued, 'the concrete link between the liberation of man and that of nature has become manifest today in the role which the ecology drive plays in the radical movement'; it reveals the way in which the domination of man takes

215

place through the domination of nature.[15] While modernity cannot return to a pre-technological stage, Marcuse argues that we must look towards 'freeing man and nature from the destructive abuse of science and technology'.[16] A recovered or liberated notion of nature could support the genuine emancipation of humanity, he suggests. In a similar vein, Mumford suggest that the technological megamachine can only be prevented from further expansion, 'with the aid of a radically different model derived directly, not from machines, but from living organisms and organic complexes (ecosystems)'.[17]

Arendt is notably sceptical about the promise of 'nature' for the human world: nature is for her not only structurally distinct from politics proper, but destructive of the fragile human artifice. Yet she is also critical of the supposed 'progress' of an increasingly wasteful consumer society, in part because this goes on 'at the expense of the world we live in'.[18] This could be read in terms of the political 'world' – and to some extent her critique of commodification is a criticism of the ephemerality of the products of modern work. But this observation is made with reference to Mumford's (insightful, she argues) critique of progress and the damage it provokes, and she continues by suggesting that 'the recent sudden awakening to the threats to our environment is the first ray of hope in this development'.[19] Although, for Arendt, natural space is not political in itself, one could suggest (drawing on similar strategies she uses elsewhere in her work) that in a state of impending crisis, such spaces can become objects of political importance or relevance.[20] The concept of natality – emphasising the centrality of life rather than death, or mortality, for political action – also brings her closer (in some ways) to some of the claims of Jonas, who prominently centres 'life' in his philosophy and environmentalism, and Marcuse, who writes that he saw the environmental and ecology movement as a means of serving 'Eros . . . the life instincts', finding 'fulfilment in the recapture and restoration of our life environment, and in the restoration of nature, both external and within human beings'.[21] Against the already partly realised, and forthcoming catastrophe of a technologised world, environmentalism offers a thread of hope for our theorists, a potential way out of the threatened moral, political, and existential collapse of modernity.

As part of his reconstruction of an ethics suitable for the challenges of technological modernity, Jonas suggested the ethical imperative might be extended to the biosphere: 'It is at least not senseless anymore to ask whether the biosphere as a whole and in its parts, now subject to our power, has become a human trust and has something of a moral claim on us not only for our ulterior sake but for its own and in its own right.'[22] He also finds in the 'dawning truth of ecology' a new and concrete opposition to the 'progressivist faith' he believes has been so prominent in technological modernity, whether socialist or capitalist.[23] To raise living standards globally, to create a degree of equality, requires a 'dizzying multiplication of planetary energy consumption', he notes. The crux of the issue of justice, therefore, is energy, and its production, management and flow, as well as its by-products and unintended consequences.[24] As the 'warning lights of various limits are coming on', he writes that ideas of utopia 'must yield to the modesty of goals that we and nature can afford'.[25] This critique of utopia is 'implicitly a critique of technology in anticipation of its extreme possibilities', possibilities that appear increasingly less extreme amid a growing awareness of contemporary environmental hazards.[26] *The Imperative of Responsibility* would go on to make an impact in the philosophy of technology. It 'deeply influenced the philosophical debates of the 1980s and 1990s', explains John-Stewart Gordon, 'and was the first major philosophical work on the dangers of ever-increasing technological progress and its (possible) negative consequences for humanity and nature'.[27] Although criticised, it was a 'politically as well as philosophically influential book' and 'has been a forerunner for the idea of sustainable development and various political programmes of socio-ecological change'.[28] It has even been claimed that Jonas is one of the most important philosophers of the twentieth century, despite having been overshadowed during his career and posthumously by others around him.[29]

Mumford has been described as the 'forgotten American environmentalist'. In the early 1970s, environmental journalist Anne Chisholm wrote 'of all the wise men whose thinking and writing over the years has prepared the ground for the environmental revolution, Lewis Mumford, the American philosopher and writer, must be preeminent'.[30] Ecological ways of thinking were built into Mumford's work, as when he argued, for instance, that modern technology's great weakness was that it possessed the 'defects of a system that, unlike organic systems, has no built-in method of controlling its growth or modulating the enormous energy it commands in order to maintain, as in any living organism, a dynamic equilibrium, favorable to life'.[31]

Ellul has also been described as 'an ardent environmentalist'.[32] His characterisation of the breakdown of modern technological society is heavily focused on its environmental impact: 'pollution . . . a congestion of clutter, an excess of information . . . the system is inundated with waste . . . technique is always on the verge of exhausting natural resources.'[33] His intellectual collaborator in the 1930s was Bernard Charbonneau, a man who late in life was recognised for his pioneering work in environmentalism. 'Ellul used to say that Charbonneau taught him to "think and be a free man"', Daniel Cérézuelle explains.[34] From the 1930s, Charbonneau 'spent all his intellectual energy to help his contemporaries realize that this uncontrolled development of industry, technology and science was the *problem* and not the solution or a neutral means for various social and political projects', working within nonconformist academic circles that were not taken seriously until the 1970s.[35] With the birth of the environmental movement, Charbonneau described how his ideas were given a 'green light', as he found himself 'suddenly found himself caught in the stream of the ecological movement'.[36] It is also notable that Ivan Illich considered Ellul to be his 'master as a social critic', another radical thinker who engages closely with environmental and ecological questions with the aim of seeking the 're-establishment of an ecological balance'.[37]

The influence of the technology critique on environmentalism

This intellectual influence between environmentalism and the critique of technology travelled in both directions. In 1978, Donald Worster noted that technological progress 'has become a target of the ecology movement'. Many ecologists, he argued, would 'contend that it is an essential part of the ecologist's job to dispel modern society's confidence in technology, and more, its faith in unlimited economic growth'.[38] Against the physicists, Worster writes, and their creation of the atomic bomb (and all the anxieties that it induced), the ecologists seemed like 'the guardians of fragile life'.[39]

In particular, an emerging understanding of the ecological significance of radioactive fallout, and growing demands in the US for more information about government weapons programs fuelled the development of the environmental movement. In 1958, scientists organised the Committee for Nuclear Information in St Louis, seeking to better understand these threats. Barry Commoner, a plant physiologist, was one of its members, and later a prominent leader in the environmentalist movement.[40] Commoner, author of the seminal 1971 work *The Closing Circle*, a work exploring the human causes of environmental damage and its relationship to modern social, economic and political systems, cites Jacques Ellul in this work. Technology 'is built on faith – in itself', writes Commoner. 'Indeed the power of technology is so evident and overwhelming as seemingly to intimidate even its critics. Thus, Jacques Ellul, one of the severest critics on effects of technology on human values, writes: 'Technique has become autonomous.'[41] Julian Huxley, the British ecologist and evolutionary biologist who served as the first director of UNESCO, a founding member of the World Wildlife Fund, and who wrote the foreword for the first British edition of *Silent Spring*, wrote in his correspondence with Mumford of how profoundly important his works, including the early *Technics and Civilization*, had been to his thought.[42] His brother Aldous, it is also worth mentioning, was responsible for the English translation of Ellul's *The Technological Society*, when he recommended it to its American publisher.

In Germany, Hans Jonas's *The Imperative of Responsibility* had a significant impact on environmental thinking. Jan C. Schmidt outlines his tangible influence on politics and legislation. Elements of Jonas's 'ethic of responsibility', built upon his critique of technology, have:

> found their way into the wording of legislation, for instance (1) the non-reciprocity dimension in the German Embryo Protection Act 1990 and

the German and Swiss Animal Protection Act ('animal as living creature', 'dignity of the creature') and (2) the caution dimension in the *Precautionary Principle* and *Risk Analysis* in the context of *Technology Assessment* methodologies . . . [while] the NGO ETC group (action group on Erosion Technology and Concentration), citing Jonas, is demanding a moratorium for nanotechnology.

Furthermore, Schmidt continues, 'Jonas has also influenced political programmes, such as the UN Commission for Environment and Development (UN General Assembly 1987). The Commission that propagated the notion of *sustainable development* was significantly inspired by Jonas . . . Jonas' imperative of responsibility is identical in content even down to the wording.'[43]

In recent years, Heidegger's work has become increasingly influential in the field of environmental theory (among other areas of contemporary political theory). This goes back further, however. 'The first deep ecologists – Arne Naess, Bill Devall, and George Sessions – all considered Heidegger to be a source of inspiration for their movement,' Casey Rentmeester writes.[44] 'Deep ecology', a term coined by Arne Naess in the early 1970s, influenced by his reading of Carson's *Silent Spring*, refers to a radical philosophy which demands the reconsideration of the place of humans as just one of many components of value in the global biosphere. It came to be a major part of the new philosophical landscape of environmental thinking and activism, as did Naess himself. What they drew from Heidegger was quite clear: 'Heidegger saw the seeds of environmental exploitation in the progress of the sciences and technological advances long before we came upon the current environmental crises of anthropogenic climate change and resource depletions in the Anthropocene.'[45] Michael Zimmerman emphasises the conceptual closeness of Heidegger's insistence on a necessary ontological shift with the shift that deep ecologists argue must necessarily take place in our understanding of the relationship between humans and nature. Yet, writes Zimmerman, it is wrong to consider Heidegger's thought itself as deep ecology due to his 'residual anthropocentrism, the reactionary dimension of his critique of industrialism', his antipathy toward science and the key fact that Heidegger was a 'severe critic of all "naturalism"'.[46] Although 'Heidegger's agrarian metaphors aimed to show humanity's relationship to the "earth"', writes Zimmerman, they also underscored 'the difference between humanity and nature'.[47] David Storey also highlights the influence of Heidegger on environmental philosophy in a different direction, citing a strand of continental philosophy known as 'ecophenomenology', which uses 'ideas of "alterity", the fullness of the human being, the importance of the phenomenon of life', as that appears in Heidegger's work.[48]

Francis Sandbach argues that a form of environmentalism he terms 'anti-establishment environmentalism' was influenced by Arendt, Ellul and Marcuse, as well as Jürgen Habermas.[49] 'In contrast to environmentalism with an eco-logical emphasis,' Sandbach argues, 'the anti-establishment movement's prin-cipal concern has been with man's alienation from society and from nature.'[50] The problem of alienation and social control, was perceived to be 'a product of science and technology', he writes. 'Within this perspective, popular ecol-ogy, systems analysis, cybernetics, decision theory, technology assessment and cost–benefit analysis could be seen as agents of social control, rather than as means of liberation.'[51] The smaller 1970s environmental movement known as the Alternative Technology movement 'was closely dependent upon this type of analysis', Sandbach writes. This movement therefore tried to define the criteria 'necessary to ensure social control over technological development. Alternative technologies would be designed so as to minimize social misuse of technol-ogy, to require few specialist skills, as well as to be non-polluting, and to use only renewable resources.'[52] Many others have also sought to draw on Arendt's ideas for ecological political theory.[53]

Understanding the world: a confluence of concepts

The conceptualisation of contemporary technology, and the 'world' produced by technologisation, connects to the contemporary environmental critique in significant ways. The concept of totality is central to the critique of technology: technology is a totalising force in politics and society, ultimately a totalitar-ian one, and its influence is globally totalising, collapsing differences between societies and political forms. The critique of technology entails a negatively unified conception of the world, the distinctions between East and West – and any other developed states or regions – are much less significant than what holds them together: the mode of technology and its political and social implications. The unity of the world, and specifically the *negative* unity of the globe, where all humans are connected by their vulnerability to a likely or even inevitable catastrophe, is just as much a part of environmental thought. In his influential essay 'The Climate of History', Dipesh Chakrabarty writes that climate change points:

> to a figure of the universal that escapes our capacity to experience the world. It is more like a universal that arises from a shared sense of a catastrophe. It calls for a global approach to politics without the myth of a global identity, for, unlike a Hegelian universal, it cannot subsume par-ticularities. We may provisionally call it a 'negative universal history'.[54]

He also observes that 'the anxiety global warming gives rise to is reminiscent of the days when many feared a global nuclear war'. Yet, the two situations have one important difference, he argues: a nuclear war would have been the result of a conscious decision, while 'climate change is an unintended consequence of human action'.[55] The critique of technology reveals that this distinction is, at least for a significant group of thinkers in the twentieth century, invalid. Technologisation itself is seen as an unconscious and autonomous force that would itself likely lead to nuclear war (or nuclear devastation by whatever means). The negative universal history arising from a shared fear of catastrophe in the Cold War world parallels that which Chakrabarty describes.

The critique of technology is a critique of progress: both a concept of progress in economic terms (because technologised productivity entails destruction), as well as the assumption of inevitable progressiveness and momentum that the critics believe structures modern liberal politics. The critique of economic progress is embedded in modern environmental movements, who similarly identify the destructive quality of unlimited production and a focus on 'productivity' at the expense of all else. Likewise, the acceleration that the critics of technology identify in modern technology, and the highly increased likelihood of risk that prevails through the consequential incapacity to predict the *effects* of technological activity, is paralleled in the acceleration of risk that environmental thinkers identify in modernity, and the unintended effects of technological human activity. Carson's *Silent Spring*, often said to have ignited the environmental movement when it was published in 1962, is about precisely the unintended, unpredictable and runaway consequences of synthetic pesticides, particularly DDT, on the natural environment.

Naturally, there are differences between the analysis of technology presented here, and the ideas contained in the early environmental movement (and *within* the discourse on technology, and *within* environmentalism, which was also made up of many discrete philosophies). The neo-Malthusianism, for example, that was so important to the critical environmentalism of the 1960s and 1970s rarely features in the work of the critics of technology, for instance.[56] The relationship between science and technology is also characterised differently by the two groups, although there is agreement on the essentially problematic character of technology itself. Early popularisers of ecological or environmental thinking saw apocalyptic dangers in the effects of runaway or poorly considered technologies but see in science a route – for some, even the primary route – to mitigating those dangers, through attempting to understand the holistic and long-term consequences of technological activity. For others, of course, it is more important to simply restrict and draw back on technological 'progress' or the utilisation of potentially risky technologies. While the critics of technology show little faith in science, the majority of (but not all) environmentalists, like the critics of technology, are characterised by a scepticism about the catastrophic consequences of unrestrained technology. Nonetheless, it is also the case that, for both groups, these claims about the use and abuse of the natural world does not necessarily mean a return to a pre-technological era, nor even necessarily a return to an 'organic' or 'natural' way of living.

Despite the existence of differences, however, the parallels are significant. In the critique of technology, we find a source of ideas for environmentalists, who encountered here a number of shared problems or questions, as well as overlapping conceptualisations of a transformed and degraded environment. This explains the mutual influence that occurs between the two schools of thought – only one of which, environmentalism, would grow spectacularly in consequence and influence. The critique of technology highlights the extraordinary power of technology and its influence on politics, economics, culture and society, but it also shows, in the words of Jacques Ellul, that 'our entire civilization', and indeed the entire world, 'is ephemeral'.[57]

CATASTROPHIC TECHNOLOGY IN THE TWENTY-FIRST CENTURY

Many of the ideas bound up in the concept of catastrophic technology thus had some influence on the environmental movements and discourses of the contemporary world. The ideas of some of the critics of technology – particularly Heidegger and Marcuse – have also been drawn on extensively in contemporary work on environmental political theory. The interest in their work comes about, in part, because of the ways in which both directly engaged with environmental questions but also because both are, in very different ways, deeply 'revolutionary' thinkers: Heidegger sought a transformation of modernity via philosophy and thought; Marcuse through a 'total' revolution in a quasi-Marxist mould. Heidegger's influence on the development of deep ecology has already been referenced, as has the ambiguous nature of the connection between his philosophy and the ideals that deep ecologists later defined. A broader connection between phenomenology and environment has also been a theme of much work. 'In contrast to the long tradition of Western philosophical dualism, phenomenology from Husserl to Heidegger and Merleau-Ponty restores attention to the human immersion in nature and the meaning immanent in ordinary experience,' Louise Westling writes.[58] Theorists have also emphasised the way in which aesthetic experience, specifically poetry, has been understood by Heidegger to be the gateway to the recovery of an authentic relationship with the environment. 'For Heidegger, poetry can, quite literally, save the earth,' observes Jonathan Bate.

> For Heidegger, language is the house of being; it is through language that unconcealment takes place for human beings. By disclosing the being of entities in language, the poet lets them be. That is the special, the sacred role of the poet. What is distinctive about the way in which humankind inhabits the earth? It is that we dwell poetically (*dichterisch*).[59]

Marcuse has been drawn on for eco-revolutionary inspiration by numerous theorists, while Anne Fremaux and John Barry point out that today's criticisms of the problems of the Anthropocene are but 'a footnote to the diagnosis outlined in Horkheimer and Adorno's *Dialectic of Enlightenment*'.[60] Contemporary political theory also draws on other aspects of the technology critique, such as that of Giorgio Agamben on the conceptualisation of biopolitics, drawing especially on Arendt and, to a lesser degree, Heidegger.

The work that most directly and fully seeks to build on the critique of technology as offered by the theorists in this book is Andrew Feenberg's critical theory of technology. This seeks to identify or recover alternatives to the dystopian technological narrative offered by the critics of technology via a more genuinely democratic, Marxist-inflected analysis that seeks to overcome the more deterministic tendencies of the technological catastrophists. Against determinism, Feenberg writes, 'the real issue is not technology or progress per se but the variety of possible technologies and paths of progress among which we must choose'. Yet, any and all alternatives 'will have political implications', he observes. 'Modern technology . . . is no more neutral than medieval cathedrals or the Great Wall of China; it embodies the values of a particular industrial civilization and especially those of elites that rest their claim to hegemony on technical mastery.' In his critical theory of technology, he seeks to 'articulate and judge these values in a cultural critique of technology . . . [and thus] begin to grasp the outlines of another possible industrial civilization based on other values'. In this, he adopts the approach of the critics of technology insofar as he argues that such a project 'requires a different sort of thinking from the dominant technological rationality, a critical rationality capable of reflecting on the larger context of technology'.[61]

Against the critics of technology (he singles out Ellul) he argues that there is 'no one single "technical phenomenon" that can be rejected as a whole'.[62] Feenberg argues that we are still free to remake technology, and to reclaim agency – he argues that, despite the dominance of technology, there is still a 'margin of maneuver' that originates within the technological system and its inevitable imperfections, breakdowns or unpredictable effects.[63] Feenberg primarily understands technology via a capitalism/socialism polarity, where socialism describes an essentially democratising force, against the capitalist-technological system. He seeks out ways to recover space for freedom against capitalism/technology and its inevitable tendency towards domination. But in the original Cold War critique of technology, the 'catastrophic' character of technology is not just its tendency to domination, nor is its nature simply (or even primarily) that of capitalism, as Feenberg suggests. Rather, to return to the critique of technology and to seek to understand it in its fullest extent offers a more multifaceted concept of technology and its influence on modernity; one whose problems are regretfully even *more* intractable, but a concept which therefore resonates more deeply with certain characteristics of the contemporary world. Two types of problem emerge in the concept of catastrophic technology that Feenberg does not deal with: first, the problem of the illusions created by technology in the modern world (the illusion of progress; the totality of the technological world as described by the idea of 'second nature'); second, and relatedly, the concrete risks associated with technology and technologies, emerging from the way in which technology invariably increases risk, including existential risk, while at the same time hiding this from plain sight under the banner of progress.

The remaining pages of this book will offer some concluding thoughts on why it might be valuable, in the twenty-first century, to return to the ideas offered by the Cold War critics of technology, and the ways in which their ideas might offer an alternative perspective from which to contemplate the challenges of technology in the modern world. The Cold War era critical technology discourse shows that people have long been thinking about the questions that seem to plague our own times so acutely: how we can deal with technology and its ill-effects, how we can understand the effect of technology on society and politics, and how we can defend ourselves against the various challenges of technology. There are clear continuities and parallels between the way that the theorists describe technology and the technologies of their own time, and those of ours, even as important specific qualities diverge – for instance, the 'democratisation' and participation enabled by social media have only reinforced the centralisation of media under the global media tech giants of today's world (along with producing its own specific social problems). The fear of a workless world – a world in which there is no more work to be undertaken, no role for the worker to play – remains now, as then, a prevalent threat, albeit still unrealised.

Against such views, there are many important and reasonable counterarguments that could be made in opposition to this conceptualisation of technology. First, that it largely dismisses the many ways that technology has improved the quality of life for billions of people: via the agricultural technologies and techniques of the Green Revolution, for instance, in which agricultural methods, machinery, fertilisers and seed varieties massively increased crop yields and productivity globally; the vaccines or antibiotics that have eradicated formerly killer illnesses or made them (relatively) benign, or the surgical techniques and anaesthetics that have made it possible to treat or cure formerly disabling or life-threatening conditions. That is to say nothing of the technologies that are to be found in households around the world: appliances that have relieved some of the laboriousness of housekeeping, mobile phones that allow access to banking for those in rural, unconnected communities across the globe, and so forth. These are not unambiguous goods, certainly, but their benefits are clear. The critics of technology are not unaware of the benefits afforded by technology. Yet they insist that 'progress' of this kind is not real progress. 'The fallen nature of modern man cannot be separated from social progress,' argue Adorno and Horkheimer, which 'allows the technical apparatus and the social groups which administer it disproportionate superiority to the rest of the population . . . the individual disappears before the apparatus which he serves, [even as] that apparatus provides for him as never before'.[64] Technology's 'goods' are always shown to be accompanied by some greater evil. It is true that the goods of the technological world are not without their problems – in some ways, the more impactful, the greater the associated problems. But the critics of technology are dogmatic in their insistence of technology's profoundly and essentially negative effect.

This results also in an analysis that is questionably deterministic. The direction of travel of technological development, and the associated catastrophic impact on the human world seems almost inexorable, although this varies between different thinkers, as has been highlighted already. But even the most 'optimistic' believe that technology's damaging effects can only be prevented or countered by radical or revolutionary change in the character of technological society. The potential for human agency is erased, and the autonomy of technology and its trajectory is dominant. This is central to the claims they make about the character of technology (and its highly problematic nature), but it is certainly not beyond critique itself.

The extent to which the critics of technology are really describing technology might also be questioned. 'Technology' merges with other political or ideological concepts and practices: rationalisation, capitalism, domination, commodification, reification and alienation, mass society, globalisation, totalitarianism, liberalism. To what extent is technology a useful or meaningful concept? To what extent is it a distinctly defined concept? Is it rather a descriptor for a more general sense of discontent in and with twentieth-century politics? Is the idea of 'technology' simply expanded beyond what we can attribute to technological artefacts and processes, or their usage?

In some ways, the concept of catastrophic technology looks like a kind of conspiracy theory. Its advocates disregard or exclude alternative evidence (the benefits of technology), identify alternative attitudes towards technology as 'illusory', and claim that technology is outside the control of those who use it, that it imposes particular behaviours on individuals who lose their agency or freedom as a result. What is missing, of course, are the conspirators – while there are some individuals or groups who might be considered particularly important (e.g. historical scientific and technological innovators such as Bacon, or more ambiguously defined capitalist elites), it is clear that for these theorists 'catastrophic technology' has become autonomous, uncontrolled and uncontrollable. This is the very problem they identify, in fact.

So is this all simply hyperbole? Much of what the critics of technology predicted has not come to pass, at least not in the ways they predicted: it does not seem *today* that we live in a world defined by nuclear power (although it did seem that way for a long time), and the fact that this has become simply one aspect of international power relations rather than the predominant concern indicates that the critics of technology were wrong about the importance of nuclear power and weapons for the modern world. For all its many and worsening problems, new and old, the world in the twenty-first century does not seem to be lurching irrevocably towards totalitarianism with the advance of technology. Globally, the make-up of liberal versus authoritarian states, parties and tendencies changes, and opinion swings from left to right and back again in different countries, but an unequivocal shift of the kind the critics of technology feared does not seem to have taken place yet. The centralisation of media critiqued by the theorists of technology has been transformed by new media technologies and the global corporations that have sprung up as a result: centralising *and* fragmenting media with unpredictable effect. We still live in a world where most people work for a living, and most people *can* work for a living – production has been automated in many cases, but work has evolved accordingly.

So what value does the narrative of catastrophic technology hold, given this litany of possible criticisms? Certainly, this idea of technology could be read as a hysterical reaction to technological development, rather than a rational, pragmatic response. But there are reasons why we might plausibly read it in a different way, as a reasonable response to certain developments in modernity. First, setting the critique in historical context is important for understanding why the theorists of technology thought what they did. They had seen the rise of totalitarianism, seen or experienced (sometimes first hand) its effects and its power: to identify modernity – a modernity that they also saw becoming increasingly bound up with technology – with totalitarianism is understandable. To attempt to seek out and illuminate aspects of or trends towards totalitarianism in modern politics and society, with a view to opposing the threat of a renewed totalitarianism, is reasonable. They read modernity through the lens of recent history. With the development, deployment and subsequent continued testing of nuclear weapons, and development of nuclear power, it also seems plausible to see in these developments the re-emergence of a spectacular, total type of domination, and an unprecedented threat, particularly amid the decades-long superpower stand-off that followed the development of the atom bomb.

Equally, in a context that remains just as relevant today, increasing awareness of the scope, scale, interconnectedness and existential risk posed by anthropogenic – technologically driven – environmental damage makes the connection between technology and catastrophe more compelling than it has ever been. The critics' analysis of nature, and their claim that in our attempts to dominate and subjugate nature, we unleash it and magnify its risks, is reflected in contemporary environmental political thought. Isabelle Stengers, for instance, describes in her work the 'intrusion of Gaia' into modernity: 'this "nature" that has left behind its traditional role and now has the power to question us all'.[65] This intrusion, she writes, 'makes a major unknown, *which is here to stay*, exist at the heart of our lives . . . no future can be foreseen in which she will give back to us the liberty of ignoring her'.[66] Stengers views this in terms of a provocation by capitalism primarily, rather than technology, but the conceptualisation of a questioning, troubling, unknown nature that humanity has unleashed upon itself can be found over half a century earlier in the work of the critics of technology. Stengers describes Gaia as a 'transcendence' that has arisen against and because of the *radically irresponsible* mode of transcendence named capitalism; for the critics of technology, out of the rationalist mode of technological modernity, arises a fundamentally irrational (unknowable, uncontrollable) nature.[67]

The critique of technology also captures, I suggest, something about the character of technological risk that is valuable to contemplate. As has been emphasised already, the character of technology is 'inhuman' – humans are shaped for technology, not technology for humans; the threat it poses is not (primarily) that of 'killer robots' which transposes a human form, motivations and control onto a mechanical device. Rather, technology, because it is increasingly autonomous, increasingly purely instrumental (or self-referential) and increasingly totalising (in the narrative offered here) poses a threat that looks very different from anything humanity has faced before, and one that is not necessarily easily recognisable *as* a threat. It is the systemic transformation of the whole world around us, and the whole world of human relationships via the mediation of technology – technology as a 'second nature'. As such, the nature of modern technology, and its consequences, cannot be understood through the models history has provided, or the ideas of ethics, politics and society that pre-dated it. One example of the way in which the critics of technology see technology transforming the spaces in which we live, and the relationships between people, is in their analysis of 'distance', as outlined in the previous chapter. Because technology increasingly makes distance meaningless, the world becomes uniform, everything is available at the same time and the same place. This stands in opposition to ideas of worldliness that rest upon dialectic, plurality or difference; it represents a global mass society. Technology, of course, *has* produced a form of global mass society (although the absolute uniformity that concerns the critics of technology has not come to pass to the extent they suggested). It is also true that this produces historically unique social constellations, such as globally diffuse but ideologically intense social networks of anonymous members. The critique of technology reminds us that the effects of technology are multitude and, if we agree that technology is fundamentally 'inhuman', we should remember that the way in which it reshapes or threatens the world is not likely to look familiar.

This is also one important way in which the critique of technology can be distinguished from a capitalist critique. Although the concept of catastrophic technology clearly incorporates central elements of anti-capitalism, from Marxist and non-Marxist positions, the significance of technological *risk* distinguishes it from other analyses of modernity that are premised upon theories of domination. It has been analysed, at length, in this study that the theorists of catastrophic technology believe domination is built into modern technology, that modern technology is a mode of domination. But it is also, because it is autonomous and purely instrumental, not a mode of domination that is wholly controlled by elites, but rather that is out of human control altogether. This is why it represents an existential risk, and what makes the theory of catastrophic technology specifically *catastrophic*. Thus, while Feenberg's critical theory of technology brilliantly analyses the critique of technology in terms of technological domination, and focuses on means of reintroducing agency and participation into technological processes, against elite domination, it excludes the question of how technology produces (and how we might mitigate) novel risks. Given that modern technology is out in the world, the answer of how to deal with technological risk does not seem to be to bring it back under human control, not least because the illusory 'progressiveness' of technology would appear to once again prioritise the furtherance of technological innovation and efficiency. The theorists of catastrophic technology make the case that technology, its risks and possibilities, must be understood in the broadest sense in terms of its structural influence on politics, society and the world.

All of this clearly presupposes that technology is not neutral. This, however, has more than one meaning. First, this book has tried to show that technology is a concept that is now fundamentally political. The idea of catastrophic technology explored in this work draws on a host of intellectual, personal and political influences and experiences; it is a conceptualisation of technology that responds to a particular time, a set of problems, a political order; it is an idea that is as much political critique as philosophical analysis or conceptual definition. This is only one way of thinking about technology, but in a world where technology is politicised, and where technology has become such a fundamental part of the fabric of politics and society, at every level, it seems unlikely that any idea of technology could now ever be politically 'neutral' (if it ever was).

The second meaning of this claim about the non-neutrality of technology relates to the nature of the thing itself, and the question of whether technology is better understood in 'instrumental' or 'substantive' terms: the former, 'the commonsense idea that technologies are "tools" standing ready to serve the purposes of their owners' versus the substantive concept, 'a minority view' which 'argues that technology constitutes a new cultural system that restructures the entire social world as an object of control.'[68] In Langdon Winner's influential 1980 article, 'Do Artifacts Have Politics?', he defends the substantive

CONCLUSION

claim, in arguing that technological artefacts can and do have political qualities of various forms. Some are flexible enough that 'their consequences for society must be understood with reference to the social actors able to influence which designs and arrangements are chosen'. In other cases, 'the intractable properties of certain kinds of technology are strongly, perhaps unavoidably, linked to particular institutionalized patterns of power and authority'.[69] The critics of technology collectively offer a similar – and even more strongly stated – defence of the claim that technologies have political qualities, both in their overarching critique of 'technology' as well as in their analysis of specific modes of technology. The idea that technologies are neutral tools, to be used for good or ill, is still, it seems fair to say, the majority or commonsense perspective. But the critics of technology offer a range of arguments and examples about the political influence of different technologies and types of technology that are worth considering in response, and which poses the question of whether there are forms of technology that should be viewed with suspicion, if we want to avoid certain political outcomes, or foster the development of particular political patterns.

Returning to the question of how we might assess the plausibility or hyperbolism of the technology critique, to what extent do their claims seem defensible, and just how catastrophic *is* technology? Certainly, there are aspects of their analysis of technology in modernity that seem to capture it with accuracy: the continual acceleration of technological development and its increasingly pervasive role in political and social life; the global and globalising character of technology; the way in which it introduces new risks into the world. There are reasonable questions that might be raised over the extent to which technology introduces new potentials for or tendencies towards control and domination (in various forms); it potentially extends the range and scope of commodification processes, although, on the other hand, there are also many ways in which technologies open up new opportunities for people to understand, engage with and experience the world. The more problematic or questionable aspects of the technology analysis are the more extreme claims: that all technology follows the same logic and tends towards the same outcomes; that the character of modern technology is domination (and inexorably leads towards totalitarian-type outcomes), and that technology is essentially catastrophic.

The degree to which technology can be characterised as essentially 'domination' and leading towards authoritarian politics is an open question. Thinkers such as those referenced here – Stengers, Feenberg, Winner (and many others) – have effectively highlighted the risks of technological domination and the need to oppose these. Certainly, technology enables greater control of populations and individuals, in ways that would not have been conceivable a hundred years ago, let alone achievable. Technology has extended and entrenched capitalism, with all the structural inequalities, subsequent limits on freedom, and political consequences that are entailed. Technology is still largely considered through an 'instrumental' lens, although at the same time (somewhat illogically) as a progressive force, although this is not, of course, universal, or without its critics. Thus, while their claims about technology are not entirely without basis, the extent to which the critics of technology make the link between totalitarian domination and technology is, I think it is fair to say, questionable.

Yet technology *does* create a new world that operates in unpredictable ways – that introduces new risks. Against this, however, we must weigh the benefits, the risks that have been controlled or eliminated: the decline in the prevalence of malnutrition; the fact that formerly common diseases have been eradicated, or brought under control; the much reduced risk of dying in childbirth, which has fallen precipitously; the capacity to medically manage pain more effectively than ever before. The critics of technology, as has been noted, were not unaware of the various elements of social progress produced by technology; they simply believed this to be relatively less important than the transformations of technological modernity on society, politics and the individual. Even where the critics of technology seemed to have the most powerful example, in the case of environmental risk, many of the outcomes they feared have not come to pass: nuclear war has so far been avoided and, while testing and nuclear accidents have had devastating local impacts in some cases, the utilisation of nuclear power does not (yet) seem to have had the effect the critics of technology feared, nor have nuclear weapons proliferated as they anticipated, thanks in part to various treaties on non-proliferation and nuclear weapons testing. Many other scenarios that early environmentalists were particularly concerned with also did not pan out as expected, notably the influence of population growth on global resources – in that case, because of the global decline in maternity rates that has resulted from the improved education, prospects for women, and contraceptive availability that has taken place over the last fifty years, as well as increased agricultural productivity. Even some of the most potentially catastrophic threats, such as the hole in the ozone layer, have been mitigated. To think of technology's effect as inherently and inevitably catastrophic, even in the case of environmental damage, is not always straightforward.

Nonetheless, the evidence of the scale and scope of the existential threats originating with industrial technology heaps up, with ever greater visibility and comprehension. And, despite this, the capacity to stop or slow the global technological-industrial behemoth is limited: consumption not only of oil but even coal is still at record highs, growing year on year. Even where we understand the risks and threat of global warming, and its causes, we lack the ability to halt those causes, continuing to 'develop' industry (including quasi-questionable 'green' technologies). In other spheres of technological development, the narrative of technological progress remains even less challenged: in medicine and biotechnologies, in digital technologies, in the development of the various gadgets and devices of the workplace and home. In the recent leap forward of artificial intelligence, there has been a rare surge of warnings and arguments put forward that its development should be halted, at least temporarily, until adequate regulation is put in place. But this is highly unusual, and the fact that it is shows how used to the assumption of technological progress we remain. The critics of technology would urge us to ask the question (whatever the answer) of whether the risks we now understand to be associated with industrial technology in the shape of global warming *and* the array of other forms of environmental damage that are being caused by industrial technology are in fact inevitable due to the character of modern technology as catastrophically risky in historically novel ways. If so, what other unknown 'side effects' might modern technology have in store for us? And does the unpredictability of the risk outweigh the many benefits that technology endows modernity with?

The critique of technology, for all its flaws, can be considered in the context of Anthropocene thought in its broader sense. First, it is broader historically: the critique of technology highlights the shortfalls of the 'sudden awakening' thesis that Bonneuil and Fressoz challenge and moves awareness of the scale of human intervention into the natural world further back. The Anthropocene is a geological concept, but it is more than that. As Chakrabarty writes, humans have become 'geological agents' but equally, 'anthropogenic explanations of climate change spell the collapse of the age-old humanist distinction between natural history and human history.'[70] The Anthropocene thesis, write Clive Hamilton, Christophe Bonneuil and François Gemenne, 'claims that humans have become a telluric force, changing the functioning of the Earth as much as volcanism, tectonics, the cyclic fluctuations of solar activity or changes in the Earth's orbital movements around the Sun'.[71] The critique of technology operates in this vein of Anthropocenic thinking, connecting and contemplating the threat of domination through technology with the wider risks of technology in the same narrative, particularly through the 'second nature' concept. Its power is in its potential to strip away the illusions of progress and to highlight the risks of technology.

It also offers something more positive. The 'grand narrative' of Anthropocenic awakening, Bonneuil and Fressoz write, as a narrative of primarily or exclusively *scientific* awakening, entails that: 'if we believe the anthropocenologist experts, serious solutions can only emerge from further technological innovation in the laboratory, rather than from alternative political experimentation "from below" in society as whole!'[72] The concept of catastrophic technology *necessitates* that we embed our thinking about technology, its risks and its future, in social and political thought. And despite the cynicism of the critics of technology (in large part) about the future of technology, to reflect on the 'catastrophe' of technology that these theorists believed was well under way by the middle of the last century, and that this catastrophe has not yet brought about the existential collapse they believed was likely, might cause us to be circumspect about the nature of the catastrophe we now face. On the other hand, it might be simply that the changes and risks identified in modernity are only the first of many, and not the last, or indeed the worst. The ambivalence of what it means to reflect on catastrophic technology only serves to highlight the necessity of staying open-minded both to threats and to the possibility of hope, and the impossibility of knowing the future in the context of such a world.

Notes

1. Duncan Bell, 'What is Liberalism?' *Political Theory* 42:6 (2014), 685.
2. Isabelle Stengers, *In Catastrophic Times: Resisting the Coming Barbarism* (London: Open Humanities Press, 2015), 29.
3. Bell, 'What is Liberalism?', 685.
4. As this book has already indicated, philosophers of technology including most significantly Langdon Winner and Andrew Feenberg have highlighted aspects of the historical narrative in their own work, but as a Cold War political narrative and philosophical position its history is still underappreciated and underexplored in wider political theory work or in the history of political thought.
5. Christophe Bonneuil and Jean-Baptiste Fressoz, *The Shock of the Anthropocene* (London: Verso, 2013), 73.
6. Ibid., 76.
7. Ibid., 148–9.
8. Ibid., 170.
9. For example: Casey Rentmeester, *Heidegger and the Environment* (Washington, DC: Rowman & Littlefield, 2015); W. S. K. Cameron, 'Heidegger's Concept of the Environment in *Being and Time*', *Environmental Philosophy* 1:1 (2004); David Macauley (ed.), *Minding Nature: The Philosophers of Ecology* (New York: The Guildford Press, 1996). among many others.
10. Murray Bookchin, 'Ecology and Revolutionary Thought' [1965], in *Post-Scarcity Anarchism* (Montreal: Black Rose Books, 1986), 44.
11. Herbert Marcuse, *Counterrevolution and Revolt* (Boston, MA: Beacon Press, 1972), 68–9.
12. Bonneuil and Fressoz, *The Shock of the Anthropocene*, 170.
13. Barbara Ward and Rene Dubos, *Only One Earth: The Care and Maintenance of a Small Planet* (Harmondsworth: Penguin, 1972), 37.
14. Marcuse, *Counterrevolution and Revolt*, 59.
15. Ibid.
16. Ibid., 60.
17. Lewis Mumford, *The Myth of the Machine: The Pentagon of Power* (New York: Harcourt Brace Jovanovich, 1970), 395–6.
18. Hannah Arendt, 'Home to Roost' [1975], in *Responsibility and Judgment* (New York: Schocken Books, 2003), 262–3.
19. Ibid., 262–3.
20. In particular, in her final work, *The Life of the Mind*, Arendt suggests that thought is in itself an *unpolitical* activity that becomes political in the modern world by virtue of its loss, such as in the case of Adolf Eichmann, who, she claimed, failed

CONCLUSION

21. Herbert Marcuse, 'Ecology and the Critique of Modern Society', *Capitalism, Nature, Socialism* 3:3 (1979), 36.

22. Jonas, 'Technology and Responsibility', 10.

23. Hans Jonas, 'Reflections on Technology, Progress and Utopia', *Social Research* 48:3 (1981), 440.

24. Jonas, *The Imperative of Responsibility*, 190.

25. Ibid., 201.

26. Ibid.

27. John-Stewart Gordon (ed.), 'Introduction', in *Global Ethics and Moral Responsibility: Hans Jonas and His Critics*, ed. John-Stewart Gordon and Holger Burckhart (Abingdon: Routledge, 2013), 22.

28. Jan C. Schmidt, 'Ethics for the Technoscientific Age: On Hans Jonas' Argumentation and His Public Philosophy Beyond Disciplinary Boundaries', in *Global Ethics and Moral Responsibility: Hans Jonas and His Critics*, ed. John-Stewart Gordon and Holger Burckhart (Abingdon: Routledge, 2013), 439.

29. David J. Levy, 'Ethics and Responsibility in a Technological Age', in *Global Ethics and Moral Responsibility: Hans Jonas and His Critics*, ed. John-Stewart Gordon and Holger Burckhart (Abingdon: Routledge, 2013).

30. Anne Chisholm, cited in Ramachandra Guha, 'Lewis Mumford, the Forgotten American Environmentalist: An Essay in Rehabilitation', in *Minding Nature: The Philosophers of Ecology*, ed. David Macauley (New York: The Guildford Press, 1996), 209.

31. Lewis Mumford, 'Science as Technology', *Proceedings of the American Philosophical Society* 105:5 (1961), 510.

32. David Lovekin, *Technique, Discourse, and Consciousness: An Introduction to the Philosophy of Jacques Ellul* (Bethlehem: Lehigh University Press, 1991), 122.

33. Jacques Ellul, 'La technique, système bloqué' [1979], cited in Lovekin, *Technique, Discourse and Consciousness*, 184.

34. Daniel Cérézuelle, 'Introduction', in Bernard Charbonneau, *The Green Light: A Self-Critique of the Ecological Movement* [1980] (London: Bloomsbury Academic, 2018), xxi.

35. Ibid., xxii–xxiii.

36. Bernard Charbonneau, *The Green Light: A Self-Critique of the Ecological Movement* [1980] (London: Bloomsbury Academic, 2018), xxxiv.

37. Ibid., 7; Ivan Illich, *Tools for Conviviality* (New York: Harper & Row, 1973), 61.

38. Donald Worster, *Nature's Economy: A History of Ecological Ideas* (Cambridge: Cambridge University Press, 1994), 22.

39. Ibid., 340.

40. Ibid., 346–7.

41. Barry Commoner, *The Closing Circle: Nature, Man and Technology* (New York: Alfred A. Knopf, 1971), 177–8.

42. Julian Huxley, letter to Lewis Mumford, 20 February 1953, folder 2338, *The Lewis Mumford Papers*, Kislak Center, University of Pennsylvania.

233

43. Schmidt, 'Ethics for the Technoscientific Age', 481–4.
44. Rentmeester, *Heidegger and the Environment*, xviii.
45. Ibid., 32.
46. Michael E. Zimmerman, *Heidegger's Confrontation with Modernity: Technology, Politics, Art* (Bloomington and Indianapolis: Indiana University Press, 1990), 243.
47. Zimmerman, *Heidegger's Confrontation with Modernity*, 196.
48. David E. Storey, *Naturalizing Heidegger* (New York: SUNY Press, 2015), 14.
49. Francis Sandbach, 'The Rise and Fall of the Limits to Growth Debate', *Social Studies of Science* 8:4 (1978), 503.
50. Ibid., 503–4.
51. Ibid.
52. Ibid.
53. For example: Kerry H. Whiteside, 'Hannah Arendt and Ecological Politics', *Environmental Ethics* 16:4 (1994); Marianne Constable, 'The Rhetoric of Sustainability: Human, All Too Human', *HA: The Journal of the Hannah Arendt Center for Politics and Humanities at Bard College* 1 (2012): 33; Anne Chapman, 'The Ways That Nature Matters: The World and the Earth in the Thought of Hannah Arendt', *Environmental Values* 16:4 (2007).
54. Dipesh Chakrabarty, 'The Climate of History: Four Theses', *Critical Enquiry* 35:2 (2009), 221–2.
55. Ibid.
56. One exception is Jonas, who was clearly very concerned about population pressures: 'We can't afford to let the population continue to grow on this planet as it has done in the last centuries and continues to do today at a statistically measurable rate. On the contrary, overpopulation – from the ecological standpoint – already represents too great a burden on the biosphere. To regulate human procreation, the political system must intervene in this most private and personal space.' Jonas, *Memoirs*, 213.
57. Ellul, *The Political Illusion*, 49–50.
58. Louise Westling, 'Merleau-Ponty's Ecophenomenology', in *Ecocritical Theory: New European Approaches*, ed. Axel Goodbody and Kate Rigby (Charlottesville: University of Virginia Press, 2011), 126.
59. Jonathan Bate, *The Song of the Earth* (London: Picador, 2000), 258.
60. For example: Charles Reitz, *Ecology and Revolution* (New York and Oxford: Routledge, 2019); Andrew Feenberg, *The Ruthless Critique of Everything Existing* (London and New York: Verso, 2023); Anne Fremaux and John Barry, 'The "Good Anthropocene" and Green Political Theory: Rethinking Environmentalism, Resisting Eco-modernism', in *Anthropocene Encounters: New Directions in Green Political Thinking*, ed. Frank Biermann and Eva Lövbrand (Cambridge: Cambridge University Press, 2019), 185.
61. Andrew Feenberg, *Transforming Technology: A Critical Theory Revisited* (Oxford: Oxford University Press, 2002), v.
62. Ibid., 14.
63. Ibid., 87.
64. Adorno and Horkheimer, *Dialectic of Enlightenment*, xiv–xv.

65. Stengers, *In Catastrophic Times*, 4.
66. Ibid., 47.
67. Ibid., 53.
68. Feenberg, *Transforming Technology*, 5–7.
69. Langdon Winner, 'Do Artifacts Have Politics?', in *The Whale and the Reactor* (Chicago: University of Chicago Press, 1986), 38.
70. Chakrabarty, 'The Climate of History: Four Theses', 206; 201.
71. Clive Hamilton, Christophe Bonneuil and François Gemenne, 'Thinking the Anthropocene', in *The Anthropocene and Global Environmental Crisis*, ed. Clive Hamilton, Christophe Bonneuil and François Gemenne (London and New York: Routledge, 2015), 3.
72. Bonneuil and Fressoz, *The Shock of the Anthropocene*, 82.

BIBLIOGRAPHY

Adorno, Theodor. *Against Epistemology: A Metacritique; Studies in Husserl and the Phenomenological Antinomies.* Cambridge, MA: MIT Press, 1986.

Adorno, Theodor. *Hegel: Three Studies.* Cambridge, MA: MIT Press, 1993.

Adorno, Theodor. *The Jargon of Authenticity.* London: Routledge and Kegan Paul, 1973.

Adorno, Theodor. 'Late Capitalism or Industrial Society?', in *Can One Live after Auschwitz?: A Philosophical Reader.* Stanford: Stanford University Press, 2003.

Adorno, Theodor. *Minima Moralia: Reflections from Damaged Life.* Translated by Edmund Jephcott. London: NLRB, 1974.

Adorno, Theodor. *Negative Dialectics* [1966]. New York: Seabury Press, 1973.

Adorno, Theodor. 'Notes on Philosophical Thinking', in *Critical Models: Interventions and Catchwords.* Translated by H. W. Pickford. New York: Columbia University Press, 1998.

Adorno, Theodor. 'Progress or Regression', in *History and Freedom: Lectures 1964–65.* Cambridge: Polity Press, 2006.

Adorno, Theodor. 'Prologue to Television', in *Critical Models: Interventions and Catchwords.* Translated by H. W. Pickford. New York: Columbia University Press, 1998.

Adorno, Theodor. 'Television as Ideology', in *Critical Models: Interventions and Catchwords.* Translated by H. W. Pickford. New York: Columbia University Press, 1998.

BIBLIOGRAPHY

Adorno, Theodor. 'Those Twenties', in *Critical Models: Interventions and Catchwords*. Translated by H. W. Pickford. New York: Columbia University Press, 1998.

Adorno, Theodor. 'Universal and Particular', in *History and Freedom: Lectures 1964–5*. Cambridge: Polity Press, 2008.

Adorno, Theodor and Horkheimer, Max. *Dialectic of Enlightenment*. London: Verso, 1997.

Agamben, Giorgio. *Remnants of Auschwitz: The Witness and the Archive*. Cambridge, MA: MIT Press, 1998.

Agamben, Giorgio. *Where are We Now? The Epidemic as Politics*. London: Urtext, 2021.

Alford, C. Fred. *Science and the Revenge of Nature: Marcuse and Habermas*. Gainesville: University Presses of Florida, 1985.

Anders, Günther. *Die Antiquiertheit des Menschen*, vol. 1: *Über die Seele im Zeitalter der zweiten industriellen Revolution*. Munich: Verlag C. H. Beck, 2010.

Anders, Günther. *Die Antiquiertheit des Menschen*, vol. 2: *Über die Zerstörung des Lebens im Zeitalter der dritten industriellen Revolution*. Munich: Verlag C. H. Beck, 2018.

Anders, Günther. 'On Promethean Shame', in *Prometheanism: Technology, Digital Culture and Human Obsolescence*. Edited by Christopher John Müller. London and New York: Rowman & Littlefield, 2016.

Anders, Günther. 'Theses for the Atomic Age', *The Massachusetts Review* 3:3 (1962), 493–505.

Arendt, Hannah. 'The Concept of History', in *Between Past and Future*. New York: Penguin, 2006.

Arendt, Hannah. 'The Crisis in Culture', in *The Promise of Politics*. New York: Schocken Books, 1993.

Arendt, Hannah. 'Europe and the Atom Bomb', in *Essays in Understanding 1930–1954: Formation, Exile, and Totalitarianism*. New York: Schocken Books, 2005.

Arendt, Hannah. 'Home to Roost', in *Responsibility and Judgment*. New York: Schocken Books, 2003.

Arendt, Hannah. *The Human Condition*. Chicago: University of Chicago Press, 1998.

Arendt, Hannah. 'The Image of Hell', in *Essays in Understanding 1930–1954: Formation, Exile, and Totalitarianism*. New York: Schocken Books, 2005.

Arendt, Hannah. 'Introduction into Politics', in *The Promise of Politics*. New York: Schocken Books, 2005.

Arendt, Hannah. 'Karl Jaspers', in *Men in Dark Times*. London: Cape, 1970.

Arendt, Hannah. 'On Violence', in *Crises of the Republic*. Harmondsworth: Penguin, 1973.

237

Arendt, Hannah. *The Origins of Totalitarianism*. Orlando: Harcourt, 1968.

Arendt, Hannah. 'What is Freedom?', in *Between Past and Future*. London: Penguin, 2006.

Aronowitz, Stanley. *Science as Power: Discourse and Ideology in Modern Society*. Minneapolis: University of Minnesota Press, 1998.

Bate, Jonathan. *The Song of the Earth*. London: Picador, 2000.

Behrent, Michael C. 'Foucault and Technology', *History and Technology* 29:1 (2013), 54–104.

Bell, Duncan. 'What is Liberalism?' *Political Theory* 42:6 (2014), 682–715.

Blok, Vincent. *Ernst Jünger's Philosophy of Technology*. New York and London: Routledge, 2017.

Bonneuil, Christophe and Fressoz, Jean-Baptiste. *The Shock of the Anthropocene: The Earth, History and Us*. London: Verso, 2013.

Bookchin, Murray. 'Ecology and Revolutionary Thought', in *Post-Scarcity Anarchism*. Montreal: Black Rose Books, 1986.

Brand, Stewart. *Whole Earth Catalog*. 1968–1972.

Brantlinger, Patrick. *Bread and Circuses: Theories of Mass Culture as Social Decay*. Ithaca: Cornell University Press, 1983.

Bronner, Stephen. *Reclaiming the Enlightenment: Toward a Politics of Radical Engagement*. New York: Columbia University Press, 2006.

Burrow, John. 'Intellectual History in English Academic Life', in *Palgrave Advances in Intellectual History*. Edited by Richard Whatmore and Brian Young. Palgrave: Basingstoke, 2006.

Cameron, W. S. K. 'Heidegger's Concept of the Environment in Being and Time', *Environmental Philosophy* 1:1 (2004), 36–46.

Canovan, Margaret. *Hannah Arendt: A Reinterpretation of Her Political Thought*. Cambridge: Cambridge University Press, 1992.

Carson, Rachel. *Silent Spring*. London: Hamish Hamilton, 1962.

Casillo, Robert. 'Lewis Mumford and the Organicist Concept in Social Thought', *Journal of the History of Ideas* 53:1 (1992), 91–116.

Cassirer, Ernst. 'Form and Technology', in *Ernst Cassirer on Form and Technology: Contemporary Readings*. Edited by Aud Sissel Hoel and Ingvild Folkvord. Basingstoke, Hampshire: Palgrave Macmillan, 2012.

Celikates, Robin and Jaeggi, Rahel. 'Technology and Reification: "Technology and Science as 'Ideology'"', in *The Habermas Handbook*. Edited by Hauke Brunkhorst, Regina Kreide and Cristina Lafont. New York: Columbia University Press, 2017.

Cérézuelle, Daniel. 'Introduction', in Bernard Charbonneau, *The Green Light: A Self-Critique of the Ecological Movement*. London: Bloomsbury Academic, 2018.

Chakrabarty, Dipesh, 'The Climate of History: Four Theses', *Critical Enquiry* 35:2 (2009), 197–222.

Chapman, Anne. 'The Ways That Nature Matters: The World and the Earth in the Thought of Hannah Arendt', *Environmental Values* 16:4 (2007), 433–55.

Charbonneau, Bernard. *The Green Light: A Self-Critique of the Ecological Movement*. London: Bloomsbury Academic, 2018.

Commoner, Barry. *The Closing Circle: Nature, Man and Technology*. New York: Alfred A. Knopf, 1971.

Constable, Marianne. 'The Rhetoric of Sustainability: Human, All Too Human', *HA: The Journal of the Hannah Arendt Center for Politics and Humanities at Bard College* 1 (2012), 158–68.

Curto, Roxanna Nydia. *Inter-tech(s): Colonialism and the Question of Technology in Francophone literature*. Charlottesville: University of Virginia Press, 2016.

Delanty, Gerard and Harris, Neal. 'Critical Theory and the Question of Technology: The Frankfurt School Revisited', *Thesis Eleven* 166:1 (2021), 88–108.

Dewey, John. *The Public and its Problems*. London: G. Allen & Unwin, 1927.

Diebold, John. *Automation*. New York: Amacom, 1983.

Edgerton, David. *The Shock of the Old: Technology and Global History since 1900*. London: Profile, 2008.

Ellul, Jacques. *Autopsy of Revolution*. Translated by Patricia Wolf. New York: Alfred A. Knopf, 1971.

Ellul, Jacques. *The Betrayal of the West*. Translated by Matthew O'Connell. New York: Seabury, 1978.

Ellul, Jacques. *The New Demons*. Translated by C. Edward Hopkin. New York: The Seabury Press, 1975.

Ellul, Jacques. *The Political Illusion*. Translated by Konrad Kellen. New York: Knopf, 1967.

Ellul, Jacques. *The Presence of the Kingdom*. Translated by Olive Wyon. Philadelphia: Westminster Press, 1951.

Ellul, Jacques. *Propaganda: The Formation of Men's Attitudes*. Translated by Konrad Kellen and Jean Lerner. New York: Knopf, 1965.

Ellul, Jacques. *The Technological Bluff*. Translated by Geoffrey Bromiley. Grand Rapids, MI: Eerdmans, 1990.

Ellul, Jacques. *The Technological Society*. Translated by John Wilkinson. New York: Random House, 1964.

Ellul, Jacques. *The Technological System*. Translated by Joachim Neugroschel. New York: Continuum, 1980.

Ellul, Jacques. *Violence: Reflections from a Christian Perspective*. Translated by Cecelia Gaul Kings. New York: Seabury, 1969.

Feenberg, Andrew 'Critical Evaluation of Heidegger and Borgmann', in *Philosophy of Technology: The Technological Condition – An Anthology*. Edited by Robert C. Scharff and Val Dusek. Oxford: Blackwell, 2003.

Feenberg, Andrew. *Heidegger and Marcuse: The Catastrophe and Redemption of History*. New York: Routledge, 2005.

Feenberg, Andrew. *The Ruthless Critique of Everything Existing*. London and New York: Verso, 2023.

Feenberg, Andrew. *Technology, Modernity and Democracy*. London and New York: Rowman & Littlefield, 2018.

Feenberg, Andrew. *Transforming Technology: A Critical Theory Revisited*. Oxford: Oxford University Press, 2002.

Ferrone, Vincenzo. *The Enlightenment: History of an Idea*. Princeton: Princeton University Press, 2015.

Firestone, Shulamith. *The Dialectic of Sex: The Case for Feminist Revolution*. New York: Bantam Books, 1971.

Forman, Paul. 'How Lewis Mumford Saw Science, and Art, and Himself', *Historical Studies in the Physical and Biological Sciences* 37:2 (2007), 271–336.

Forman, Paul. 'The Primacy of Science in Modernity, of Technology in Postmodernity, and of Ideology in the History of Technology', *History and Technology* 23:1/2 (2007), 1–152.

Francastel, Pierre. *Art and Technology in the Nineteenth and Twentieth Centuries*. New York: Zone Books, 2000.

Fremaux, Anne and Barry, John. 'The "Good Anthropocene" and Green Political Theory: Rethinking Environmentalism, Resisting Eco-modernism', in *Anthropocene Encounters: New Directions in Green Political Thinking*. Edited by Frank Biermann and Eva Lövbrand. Cambridge, Cambridge University Press, 2019.

Friedmann, Georges. *The Anatomy of Work*. London: Heinemann Educational Books, 1961.

Fuller, Buckminster. *Operating Manual for Spaceship Earth*. Carbondale: Southern Illinois University Press, 1969.

Gordon, John-Stewart (ed.). 'Introduction', in *Global Ethics and Moral Responsibility: Hans Jonas and His Critics*. Edited by John-Stewart Gordon and Holger Burckhart. Abingdon: Routledge, 2013.

Grant, George. 'In Defence of North America', in *Technology and Empire*. Toronto: House of Anansi Press, 1969.

Grant, George. 'Thinking about Technology', *Communio* 28 (2001), 11–34.

Guha, Ramachandra. 'Lewis Mumford, the Forgotten American Environmentalist: An Essay in Rehabilitation', in *Minding Nature: The Philosophers of Ecology*. Edited by David Macauley. New York: The Guildford Press, 1996.

Hamilton, Clive, Bonneuil, Christophe and Gemenne, François. 'Thinking the Anthropocene', in *The Anthropocene and Global Environmental Crisis*. Edited by Clive Hamilton, Christophe Bonneuil and François Gemenne. London and New York: Routledge, 2015.

BIBLIOGRAPHY

Heidegger, Martin. 'The Age of the World Picture', in *The Question Concerning Technology and Other Essays*. Translated by William Lovitt. New York: Harper & Row, 1977.

Heidegger, Martin. 'Anaximander's Saying', in *Off the Beaten Track*. Translated by Julian Young and Kenneth Haynes. Cambridge: Cambridge University Press, 2002.

Heidegger, Martin. *Four Seminars*. Bloomington and Indianapolis: Indiana University Press, 2003.

Heidegger, Martin. 'Insight into That Which Is', in *The Bremen and Freiburg Lectures*. Translated by Andrew J. Mitchell. Bloomington and Indianapolis: Indiana University Press, 2012.

Heidegger, Martin. 'The Origin of the Work of Art', in *Off the Beaten Track*. Translated by Julian Young and Kenneth Haynes. Cambridge: Cambridge University Press, 2002.

Heidegger, Martin. *Parmenides*. Bloomington and Indianapolis: Indiana University Press, 1992.

Heidegger, Martin. *Ponderings XII–XV*. Translated by Richard Rojcewicz. Bloomington and Indianapolis: Indiana University Press, 2017.

Heidegger, Martin. 'The Question Concerning Technology', in *The Question Concerning Technology and Other Essays*. Translated by William Lovitt. New York: Harper & Row, 1977.

Heidegger, Martin. 'The Thing', in *Poetry, Language, Thought*. Translated by Albert Hofstadter. Harper & Row: New York, 1971.

Heidegger, Martin. *What is a Thing?* Translated by W. B. Barton Jr. and Vera Deutsch. Chicago: Henry Regnery Company, 1967.

Heidegger, Martin. 'Why Poets?', in *Off the Beaten Track*. Translated by Julian Young and Kenneth Haynes. Cambridge: Cambridge University Press, 2002.

Herf, Jeffrey. '"Dialectic of Enlightenment" Reconsidered', *New German Critique* 117 (2012), 81–9.

Herf, Jeffrey. *Reactionary Modernism: Technology, Culture, and Politics in Weimar and the Third Reich*. Cambridge: Cambridge University Press, 1984.

Horkheimer, Max. 'The Concept of Man', in *Critique of Instrumental Reason*. London: Verso, 2012.

Horkheimer, Max. *Dawn and Decline*. New York: The Seabury Press, 1978.

Horkheimer, Max. *Eclipse of Reason*. London: Continuum, 1947.

Horkheimer, Max. 'Foreword', in *Critique of Instrumental Reason*. London: Verso, 2012.

Horkheimer, Max. 'Threats to Freedom', in *Critique of Instrumental Reason*. London: Verso, 2012.

Hughes, Thomas P. *Human-Built World: How to Think about Technology and Culture*. Chicago: University of Chicago Press, 2004.

Husserl, Edmund. *The Crisis of European Sciences and Transcendental Phenomenology*. Evanston: Northwestern University Press, 1970.

Huxley, Julian. letter to Lewis Mumford, 20 February 1953. In Folder 2338, *The Lewis Mumford Papers*. Kislak Center, University of Pennsylvania.

Ihde, Don. *Heidegger's Technology: Postphenomenological Perspectives*. New York: Fordham University Press, 2010.

Ihde, Don. 'A Phenomenology of Technics,' In *Philosophy of Technology: The Technological Condition – An Anthology*. Edited by Robert C. Scharff and Val Dusek. Oxford: Blackwell, 2003.

Illich, Ivan. *Tools for Conviviality*. New York: Harper & Row, 1973.

Jay, Martin. *The Dialectical Imagination: A History of the Frankfurt School and the Institute of Social Research, 1923–1950*. Berkeley: The University of California Press, 1973.

Jenemann, David. *Adorno in America*. Minneapolis: University of Minnesota Press, 2007.

Jonas, Hans. 'Acting, Knowing, Thinking: Gleanings from Hannah Arendt's Philosophical Work', *Social Research* 44:1 (1977).

Jonas, Hans. 'Against the Stream: Comments on the Definition and Redefinition of Death', in *Philosophical Essays*. New York and Dresden: Atropos Press, 1980.

Jonas, Hans. 'Biological Engineering – A Preview', in *Philosophical Essays*. New York and Dresden: Atropos Press, 1980.

Jonas, Hans. 'Biological Foundations of Individuality', in *Philosophical Essays*. New York and Dresden: Atropos Press, 2010.

Jonas, Hans. 'Contemporary Problems in Ethics from a Jewish Perspective', in *Philosophical Essays*. New York and Dresden: Atropos Press, 1980.

Jonas, Hans. *The Imperative of Responsibility*. Chicago: University of Chicago Press, 1984.

Jonas, Hans. 'Introduction', in *Philosophical Essays*. New York and Dresden: Atropos Press, 1980.

Jonas, Hans. *Memoirs: Hans Jonas*. Waltham: University Press of New England, 2008.

Jonas, Hans. *The Phenomenon of Life*. New York: Harper & Row, 1966.

Jonas, Hans. 'Philosophical Reflections on Experimenting with Human Subjects', in *Philosophical Essays*. New York and Dresden: Atropos Press, 1980.

Jonas, Hans. 'Prologue', in *Mortality and Morality: A Search for Good after Auschwitz*. Edited by Lawrence Vogel. Evanston: Northwestern University Press, 1996.

Jonas, Hans. 'Reflections on Technology, Progress and Utopia', *Social Research* 48:3 (1981), 411–55.

Jonas, Hans. 'Seventeenth Century and After: The Meaning of Scientific and Technological Revolution', in *Philosophical Essays*. New York and Dresden: Atropos Press, 1980.

Jonas, Hans. 'Technology and Responsibility: Reflections on the New Tasks of Ethics', in *Philosophical Essays*. New York and Dresden: Atropos Press, 1980.

Jonas, Hans. 'Technology as a Subject for Ethics', *Social Research* 49:4 (1982), 891–8.

Jonas, Hans. 'Toward a Philosophy of Technology', *The Hastings Centre Report* 9:1 (1979), 34–43.

Jonas, Hans. 'Towards an Ontological Grounding', in *Mortality and Morality: A Search for Good after Auschwitz*. Edited by Lawrence Vogel. Evanston: Northwestern University Press, 1996.

Jünger, Ernst. *The Worker*. Evanston: Northwestern University Press, 2017.

Kelly, Duncan. *Politics and the Anthropocene*. Cambridge: Polity, 2019.

Koselleck, Reinhart. *Crisis and Critique: Enlightenment and the Modern Pathogenesis of Modern Society*. Cambridge, MA: The MIT Press, 1988.

Lazier, Benjamin. 'Earthrise; or, The Globalization of the World Picture', *The American Historical Review* 116:3 (2011).

Levy, David J., 'Ethics and Responsibility in a Technological Age', in *Global Ethics and Moral Responsibility: Hans Jonas and His Critics*. Edited by John-Stewart Gordon and Holger Burckhart. Abingdon: Routledge, 2013.

Lovekin, David. *Technique, Discourse, and Consciousness: An Introduction to the Philosophy of Jacques Ellul*. Bethlehem: Lehigh University Press, 1991.

Lukács, Georg. 'Reification and the Consciousness of the Proletariat', in *History and Class Consciousness*. Cambridge, MA: The MIT Press, 1971.

Macauley, David (ed.). *Minding Nature: The Philosophers of Ecology*. New York: The Guildford Press, 1996.

Marcuse, Herbert. 'Aggressiveness in Advanced Industrial Society', in *Negations: Essays in Critical Theory*. Translated by Jeremy J. Shapiro. London: Free Association, 1988.

Marcuse, Herbert. *Counterrevolution and Revolt*. Boston, MA: Beacon Press, 1972.

Marcuse, Herbert. 'Ecology and the Critique of Modern Society', *Capitalism, Nature, Socialism* 3:3 (1979), 29–38.

Marcuse, Herbert. 'The End of Utopia', in *Five Lectures*. Boston, MA: Beacon, 1970.

Marcuse, Herbert. *Eros and Civilization: A Philosophical Inquiry into Freud*. London: Ark, 1987.

Marcuse, Herbert. *An Essay on Liberation*. Boston, MA: Beacon Press, 1969.

Marcuse, Herbert. 'Industrialization and Capitalism in the Work of Max

Weber', in *Negations: Essays in Critical Theory*. London: MayFlyBooks, 1968.

Marcuse, Herbert. 'The Inner Logic of American Policy in Vietnam', in *Teach-Ins, USA: Reports, Opinions, Documents*. Edited by Louis Menashe and Ronald Radosh. New York: Praeger, 1967.

Marcuse, Herbert. 'Liberation from the Affluent Society', in *The Dialectics of Liberation*. Edited by David Cooper. Harmondsworth and Baltimore: Penguin, 1968.

Marcuse, Herbert. 'Marxism and the New Humanity: An Unfinished Revolution', in *Marxism and Radical Religion: Essays Toward a Revolutionary Humanism*. Edited by John C. Raines and Thomas Dean. Philadelphia: Temple University Press, 1970.

Marcuse, Herbert. *One-Dimensional Man*. Boston, MA: Beacon, 1964.

Marcuse, Herbert. 'Socialist Humanism', in *Socialist Humanism: An International Symposium*. Edited by Erich Fromm. Garden City: Doubleday, 1965.

Marcuse, Herbert. 'Some Social Implications of Modern Technology', *Studies in Philosophy and Social Science* 9 (1941), 414–39.

Marcuse, Herbert. *Soviet Marxism: A Critical Analysis*. London: Routledge and Kegan Paul, 1958.

Marcuse, Herbert. 'World without a Logos', *Bulletin of the Atomic Scientists* 20 (January 1964), 25–6.

Marlin, Randal. 'Jacques Ellul and the Nature of Propaganda in the Media', in *The Handbook of Media and Mass Communication Theory*. Chichester: John Wiley & Sons, 2014.

Marx, Leo. '"Technology": The Emergence of a Hazardous Concept', *Social Research* 64:3 (1997), 965–88.

McDermott, John. 'Technology: The Opiate of the Intellectuals', in *Technology and Man's Future*. Edited by Albert H. Teich. New York: St. Martin's Press, 1981.

Melzer, Arthur M., Weinberger, Jerry and Zinman, M. Richard (eds). 'Preface', in *Technology in the Western Political Tradition*. Ithaca and London: Cornell University Press, 1993.

Mesthene, Emanuel. 'Technology and Wisdom', in *Philosophy of Technology: The Technological Condition – An Anthology*. Edited by Robert C. Scharff and Val Dusek. Oxford: Blackwell, 2003.

Miller, Tyrus. *Modernism and the Frankfurt School*. Edinburgh: Edinburgh University Press, 2014.

Mitcham, Carl. *Thinking through Technology: The Path between Engineering and Philosophy*. Chicago: University of Chicago Press, 1994.

Moyn, Samuel. *Liberalism against Itself: Cold War Intellectuals and the Making of Our Times*. New Haven and London: Yale University Press, 2023.

Müller, Christopher John. 'Introduction', in *Prometheanism: Technology, Digital Culture and Human Obsolescence*. London and New York: Rowman & Littlefield, 2016.

Mumford, Lewis. *The Conduct of Life*. New York: Harcourt, Brace & Co., 1951.

Mumford, Lewis. 'The Corruption of Liberalism', *The New Republic*, 29 April 1940.

Mumford, Lewis. *The Myth of the Machine: The Pentagon of Power*. New York: Harcourt Brace Jovanovich, 1970.

Mumford, Lewis. *The Myth of the Machine: Technics and Human Development*. New York: Harcourt Brace Jovanovich, 1967.

Mumford, Lewis. *My Works and Days: A Personal Chronicle*. New York and London: Harcourt Brace Jovanovich, 1979.

Mumford, Lewis. 'Science as Technology', *Proceedings of the American Philosophical Society* 105:5 (1961), 506–11.

Mumford, Lewis. *Technics and Civilization*. San Diego, New York and London: Harcourt Brace Jovanovich, 1963.

Mumford, Lewis. 'Technics and the Nature of Man', *Technology and Culture* 7:3 (1966), 303–17.

Omachonu, John O. and Healey, Kevin. 'Media Concentration and Minority Ownership: The Intersection of Ellul and Habermas', *Journal of Mass Media Ethics* 24:2/3 (2009), 90–109.

Pitkin, Hanna F. 'Rethinking Reification', *Theory and Society* 16:2 (1987), 263–93.

Rabinbach, Anson. *In the Shadow of Catastrophe: German Intellectuals between Apocalypse and Enlightenment*. Berkeley: University of California Press, 1997.

Reitz, Charles. *Ecology and Revolution*. New York and Oxford: Routledge, 2019.

Rentmeester. Casey, *Heidegger and the Environment*. Washington, DC: Rowman & Littlefield, 2015.

Rorvik, David. *Brave New Baby*. London: New English Library, 1978.

Sandbach, Francis. 'The Rise and Fall of the Limits to Growth Debate', *Social Studies of Science* 8:4 (1978), 495–520.

Schatzberg, Eric. *Technology: Critical History of a Concept*. Chicago: University of Chicago Press, 2018.

Schell, Jonathan. 'In Search of a Miracle: Hannah Arendt and the Atomic Bomb', in *Politics in Dark Times*. Edited by Seyla Benhabib. Cambridge: Cambridge University Press, 2010.

Schmidt, Jan C. 'Ethics for the Technoscientific Age: On Hans Jonas' Argumentation and His Public Philosophy beyond Disciplinary Boundaries', in *Global Ethics and Moral Responsibility: Hans Jonas and His Critics*. Edited by John-Stewart Gordon and Holger Burckhart. Abingdon: Routledge, 2013.

Sheehan, Stephen. 'The Nature of Technology: Changing Concepts of Technology in the Early Twentieth-Century', *Icon* 11 (2005), 1–15.

Simbirski, Brian. 'Cybernetic Muse: Hannah Arendt on Automation, 1951–1958', *Journal of the History of Ideas* 77:4 (2017), 589–613.

Spengler, Oswald. *Man and Technics: A Contribution to a Philosophy of Life.* Budapest: Arktos, 2015.

Stengers, Isabelle. *In Catastrophic Times: Resisting the Coming Barbarism.* Open Humanities Press, 2015.

Storey, David E. *Naturalizing Heidegger.* New York: SUNY Press, 2015.

Teich, Albert H. (ed). 'Introduction', in *Technology and Man's Future.* New York: St. Martin's Press, 1981.

Tschachler, Heinz. '"Hitler Nevertheless Won the War": Lewis Mumford's Germany and American Idealism', *American Studies* 44:1 (1999).

Vázquez-Arroyo, Antonio Y. 'How Not to Learn from Catastrophe: Habermas, Critical Theory, and the "Catastrophization" of Political Life', *Political Theory* 41:5 (2013), 738–65.

Villa, Dana. *Arendt and Heidegger: The Fate of the Political.* Princeton: Princeton University Press, 1996.

Villa, Dana. *Public Freedom.* Princeton: Princeton University Press, 2008.

Vogel, Lawrence. 'Introduction', in *Mortality and Morality: A Search for Good after Auschwitz.* Edited by Lawrence Vogel. Evanston: Northwestern University Press, 1996.

Vogel, Steven, 'On Nature and Alienation', in *Critical Ecologies: The Frankfurt School and Contemporary Environmental Crises.* Edited by Andrew Biro. Toronto: University of Toronto Press, 2011.

Ward, Barbara and Dubos, René, *Only One Earth: The Care and Maintenance of a Small Planet.* Harmondsworth: Penguin, 1972.

Weinberger, Jerry, 'Liberal Democracy and the Problem of Technology', in *Democratic Theory and Technological Society.* Edited by Richard B. Day, Ronald Beiner and Joseph Masciulli. Abingdon: Routledge, 1988.

Westling, Louise, 'Merleau-Ponty's Ecophenomenology', in *Ecocritical Theory: New European Approaches.* Edited by Axel Goodbody and Kate Rigby. Charlottesville: University of Virginia Press, 2011.

Whiteside, Kerry H. 'Hannah Arendt and Ecological Politics', *Environmental Ethics* 16:4 (1994), 339–58.

Wiggershaus, Rolf. *The Frankfurt School.* Cambridge, MA: MIT Press, 1994.

Wilson, Harold. 'Labour and the Scientific Revolution', Policy statement made to the Annual Conference of the Labour Party, Scarborough, 1963.

Winner, Langdon. *Autonomous Technology: Technics-out-of-Control as a Theme in Political Thought.* Cambridge, MA: The MIT Press, 1977.

Winner, Langdon. 'Do Artifacts Have Politics?', in *The Whale and the Reactor.* Chicago: University of Chicago Press, 1986.

Winner, Langdon. 'Technologies as Forms of Life', in *The Whale and the Reactor*. Chicago: University of Chicago Press, 1986.

Wolin, Sheldon. *Fugitive Democracy*. Princeton: Princeton University Press, 2016.

Worster, Donald. *Nature's Economy: A History of Ecological Ideas*. Cambridge: Cambridge University Press, 1994.

Yaqoob, Waseem. 'The Archimedean Point: Science and Technology in the Thought of Hannah Arendt, 1951–1963', *Journal of European Studies* 44:3 (2014), 199–224.

Zimmerman, Michael E. *Heidegger's Confrontation with Modernity: Technology, Politics, Art*. Bloomington and Indianapolis: Indiana University Press, 1990.

INDEX

acceleration, 81–2, 102–5, 123–4, 202–3,
  221
Adorno, Theodor, 25–7, 33–6, 84–6,
  152–9
  against Heidegger, 43–4
  alienation, 109–10, 123, 126–32
  culture, 45, 84–6, 143–6, 148–9, 195
  myth, 65
  progress, 199–202, 224–5
  see also Dialectic of Enlightenment,
  culture industry
aesthetics, 26, 34, 45, 47, 130, 135, 189,
  222; see also art
agriculture, 51, 102, 106
  pesticides, 103, 221
alienation, 149, 166, 194–6, 220
  earth alienation, 41, 52–3, 69, 193, 196
  of power, 109
  production, 119–20, 123–6, 128–31,
  135
  self-alienation, 51, 126, 131
  world alienation, 41, 193
  see also Lukács, Georg, reification
Alternative Technology movement, 220
America, 37, 39, 53, 158, 199–200
  Cold War, 5, 96, 111–13, 211

German immigration, 4–5, 13–14, 33–4,
  143
war, 29–30, 49, 106
Anders, Günther, 23, 25–7, 37–9, 41,
  47–8, 52, 192, 194–8
  freedom, 145
  media, 150–4
  nuclear power, 82–4, 100–2
  production, 125–7
  raw material, 166, 179–180, 182, 202
  reification, 131–3
  tools, 121
  totalitarianism, 107, 109, 114, 120, 157–8
Anthropocene, 9, 212–14, 219, 222–3,
  231–2
anti-establishment environmentalism,
  219–20
apocalypse, 1, 8, 16, 19, 36, 39–40, 108–9,
  203
  nuclear, 18, 50, 102–5
Arendt, Hannah, 15, 24–7, 37–41, 46–8,
  113–14
  alienation, 70, 196–7
  atom bomb, 18, 80, 100–2, 105, 107–11
  automation, 52, 74
  the body, 165–72, 177, 179, 182

248

INDEX

capitalism, 76–7
Christianity, 67, 169–70
freedom, 145
Heidegger, 43–4, 181
labor, 120–7, 132–4, 136
nature, 101, 187, 190–3, 201–2, 216
progress, 200
public and private spheres, 146–8, 151–2
science, 69–71, 88
space, 53
Technê, 65–6
totalitarianism, 81–4, 98–9, 107–111, 153
art, 47, 65, 106, 135; *see also* aesthetics
artifice
political, 41, 81, 102, 104, 132, 168–9, 171, 190, 193, 202, 216
technological, 53, 88, 132, 186
automation, 19, 51–2, 74, 77–8, 80, 82, 118–27, 130, 132–7, 175, 180
automobiles, 36, 128, 134

Bacon, Francis, 51, 68–9, 71, 175
barbarism, 30, 49, 75, 84, 105, 109, 112–13, 179
being, 40, 76, 83, 89, 105, 151, 173–5
Heidegger, 31, 43–4, 83–4, 110, 133–4, 181–2
Bell, Duncan, 5, 211
Benjamin, Walter, 19, 27, 39, 130, 135, 146
biotechnology, 165, 172–7
the body, 19–20, 99, 106, 122, 126, 163–82
Bonneuil, Christophe, 212–14, 231
Bookchin, Murray, 214
bureaucracy, 18, 29, 35, 97, 119, 145, 157, 168

capitalism, 45, 47, 76–8, 119–21, 124–30, 143, 227
end of capitalism, 123, 159, 223
freedom, 135, 158
ideology, 51–2, 73
Carson, Rachel, 9, 29, 213, 219, 221
catastrophic technology, 1–8, 17–19, 24, 28, 107, 211–12, 223–8, 231–2
Chakrabarty, Dipesh, 220–1, 231
Charbonneau, Bernard, 30, 194, 217

Christianity, 67–68, 86, 166, 168–71, 181
Ellul, 26, 30, 32–3, 46, 48
Gnosticism, 38–9, 63
civilisation, 49, 79, 110, 133, 190, 199
civilisational collapse, 30, 42, 101, 106
Cold War, 4–6, 8–10, 15–17, 26–8, 50, 110–12, 212
commodification, 129, 131, 192, 216, 229
Commoner, Barry, 9, 213, 218
communism, 5–6, 8–9, 16–17, 35, 39, 51, 84–5, 110–11
concentration camps, 48, 108–9, 166–8
critical theory of technology, 7, 223–4, 228
culture industry, 36, 45, 85–6, 143, 148–52
cybernetics, 16, 52, 220

Darwin, Charles, 170, 175
deep ecology, 219, 222
dehumanisation, 18, 31–2, 36, 41, 52–4, 66, 87, 106, 133, 166
democracy, 17, 34, 110–11, 144–5
illusion, 85, 157–8
Descartes, René, 70–1, 173–5, 179;
see also dualism
dialectic, 18, 32, 44–6, 53–4, 73, 198, 227
freedom, 145, 152, 156–7
tension, 147–8, 158
Dialectic of Enlightenment, 23, 50–1, 69, 73–5, 113
distance, 151, 197–8, 227
domination, 54, 69, 73–4, 79–80, 108, 122–3, 156–9, 195, 228–9
consumption, 127–8, 134, 144, 146
of humans, 9, 18, 35–6, 75, 83, 165, 201, 215
media, 85–6
nature, 9, 35–6, 50–1, 75, 81, 83, 190, 201, 215
dualism, 71, 173–4, 181, 222

earth, 205, 231
earth system, 212
earthliness, 191, 196
Arendt, 44, 52–3, 70, 80, 101–2, 111, 169, 197–8
Heidegger, 181, 188, 219, 222
standing-reserve, 43, 51, 76
see also alienation: earth alienation, world

249

Earthrise, 9, 193, 198–9
economics, 36, 121, 127, 156, 221
  Economic Man, 133
  see also capitalism, Marxism
ecophenomenology, 219
efficiency, 18, 21n, 79–80, 109, 200
  megamachine, 66
  technique, 32, 51
Ellul, Jacques, 7, 25–33, 42, 48–9, 218
  atom bomb, 99–100, 102
  Christianity, 32–3, 67, 86
  democracy, 144–5, 158
  dialectic, 198
  Enlightenment, 75, 78–9, 154
  environment, 205, 217–19, 222
  globalisation, 111, 199
  Lenin, 85
  liberalism, 4
  the machine, 52, 78
  Marx, 44, 46
  nature, 188, 190–2, 194–6, 205
  philosophy of technology, 7, 24
  pollution, 103
  private sphere, 147–8, 150
  progress, 200–1
  propaganda, 53, 153–7
  science, 72
  technique, 9, 32, 51, 65–7, 178–9, 203
  totalitarianism, 107, 109–11, 157
  utopia, 88
  war, 98, 112
  work, 120, 124, 127, 137
  see also technique
energy, 88, 103, 122, 132, 182, 190,
  201–2, 216–17
  nuclear energy, 40, 51, 70, 76, 80, 99,
  101, 215
enlightenment, 42, 51, 69, 154
  the Enlightenment, 35–6, 45, 49, 73–80,
  113, 126
  nature, 130, 188, 190
environment, 65, 72, 186–96, 230
environmentalism, 7, 9–10, 20, 212–22
  catastrophe, 3, 37, 98, 102–3, 202, 205,
  226
  nature, 71, 82, 180, 204
ethics, 39–41, 172–6, 181, 216
  ethical collapse, 51, 83, 97, 103, 175–6,
  189
  ethic of responsibility, 54–5, 89, 218–19

fascism, 44, 51, 85, 99, 105, 112, 155
  German, 30, 33, 49, 166
Feenberg, Andrew, 7–8, 47, 68, 134–5,
  223–4, 228
film, 36, 142, 151, 155, 198
First World War, 42, 54, 84–5
Foucault, Michel, 7, 164, 176
France, 5, 11, 42, 156, 164; *see also*
  French Revolution
Frankfurt School, 3, 23–7, 33–9, 43–5, 50,
  79, 120, 130, 143
  Zeitschrift für Sozialforschung, 34, 39
freedom, 54, 75, 97–8, 107–9, 144–6,
  155–9, 200–1, 204
  necessity, 45–6, 127, 132–5
  political freedom, 101–2
  spontaneity, 47, 85, 108, 123, 145,
  151–2, 178
French Revolution, 54, 64, 75, 78
Fressoz, Jean-Baptiste, 212–14, 231
Friedmann, Georges, 118, 132

Galileo, 52, 54, 68–9, 71–2, 154
genetics, 1, 8, 18, 88, 163, 175–6
genocide, 36, 105, 113
Germany, 13, 24, 34–5, 38–9, 48–50, 84,
  106, 218
  German Revolution, 34, 39, 46
  Nazi Germany, 14, 27, 30, 35, 98, 113,
  158, 166–7
  Weimar Germany, 5, 14, 30, 39, 42–3,
  49, 55
globalisation, 8, 110, 113, 198–9, 225, 229
Gnosticism, 38–9, 63
Grant, George, 6, 200
Greece, 11, 39, 47, 63–7, 70, 86, 106, 147,
  178, 190, 193; s*ee also* technê

Habermas, Jürgen, 7, 130, 220
Hegel, G. W. F., 73, 99, 220
Heidegger, Martin, 2, 7, 23–9, 51, 136–7,
  200
  critiques of, 37, 41, 173–5, 181–2, 206n
  enframing, 31, 70, 76–7, 88, 193
  Greece, 64–7, 70–1
  influence, 3–4, 13–14, 23–4, 35, 38–9,
  42–4, 47, 52, 219
  machination, 109–10, 181, 204
  nature, 188–9, 191–3, 198, 202, 213–14,
  222

250

INDEX

Nazism, 48–9, 84, 105–6
nuclear, 83, 100–4
standing-reserve, 31, 43, 76–7, 120, 133, 151, 202
Herf, Jeffrey, 42–3, 106, 130
holocaust, 28, 36, 41, 49–50, 84, 105
    atomic holocaust, 97, 102
    Auschwitz, 36, 38, 99, 133, 166
Horkheimer, Max, 24–7, 33–6, 47–8, 63, 88
    alienation, 109, 130–2
    capitalism, 126–30
    democracy, 157–9
    enlightenment, 4, 113
    freedom, 145
    the Frankfurt School, 44–5
    mass media, 153
    nature, 188, 190–1, 202
    public sphere, 148–9, 152
    reason, 65, 69, 73, 75, 79–80, 85, 110, 153–5, 165, 201
    self, 178, 182, 194–5, 224–5
    see also *Dialectic of Enlightenment*, culture industry
human nature, 52, 74, 102–4, 133, 173, 186–8, 194, 201; see also the body
Husserl, Edmund, 39, 48, 68–9, 222

Ihde, Don, 3–4, 49–50, 106, 195
The Imperative of Responsibility, 24, 38, 40, 44, 54, 89, 176, 180, 217–19
industrialisation, 118, 171, 178, 188, 193
    Industrial Revolution, 76–80, 100–1, 121–2, 126
    industrial society, 53, 82, 123, 128, 130, 150, 195, 213–14, 223
    industrial technologies, 31, 49, 102, 105, 230–1
instrumentalisation, 41, 65, 74–6, 79, 87, 108, 198, 203–4, 228

Jaspers, Karl, 38, 43
Jay, Martin, 34, 44–5, 50, 143
Jonas, Hans, 23–7, 38–48, 71, 88, 108, 125, 197, 202, 216
    the body, 165, 172–82
    environmentalism, 218–19
    existential risk, 8, 18, 54–5, 83, 102–3, 202–4
    Industrial Revolution, 76, 121
    nature, 98, 189–90, 192, 199

nuclear technology, 82–3
technê, 65–6
see also *The Imperative of Responsibility*
Jünger, Ernst, 10, 16, 25, 42–3, 49, 99, 106
Jünger, Friedrich, 42

Kant, Immanuel, 25, 47, 137, 174
Kierkegaard, Søren, 25, 30, 47
Koselleck, Reinhart, 24, 200

labour, 41, 66, 74, 118–37, 154, 166, 170, 187
    alienation, 119–20, 124, 169, 178
    animal laborans, 77, 124, 133
    division of labour, 51, 126
    see also Arendt, Hannah, automation, production
Lazier, Benjamin, 9, 187, 191, 193, 198–9
leisure, 41, 124, 127–8, 136, 163
liberalism, 4–6, 15, 26–8, 31, 75, 84–5, 113, 144, 158, 211–12
    liberal democracy, 8, 17, 50, 84, 110, 144
    progress, 15–16, 200
life, 67, 169–71, 182
    biological life, 165–6, 170–1
    life process, 74, 77, 126, 146–7, 177
    metabolism, 170, 174–5, 179, 189–90
Lukács, Georg, 7, 42, 128–30, 192; see also reification

Marcuse, Herbert, 7, 25–7, 33–7, 43, 47–8, 53, 197
    aesthetics, 47, 135
    alienation, 128–32, 135, 195, 202
    consumption, 127, 149–50
    environmentalism, 213–16, 222–3
    freedom, 145, 200–1
    human nature, 103, 180, 190–1
    instrumentality, 74–5, 108–9, 178, 203
    Marx, 44–6, 119–21, 134–5
    nature, 192, 204
    optimism, 88, 134, 136
    science, 72–3
    Soviet Union, 17, 35–6, 157–8
    standardisation, 123, 157
    totalitarianism, 16–17, 74, 82
    see also reification, second nature
Marx, Leo, 7

251

Marxism, 25–7, 33–9, 44–6, 87–8, 119–21, 134, 143; *see also* Frankfurt School, reification
mechanisation, 41, 54, 123, 127, 129–32
media, 34, 53, 142–59, 195, 224, 226; *see also* culture industry, film, propaganda, radio, television
mediation, 9, 33, 79, 129, 137, 143, 149, 153, 158–9, 194–6, 227
megamachine, 31–2, 54, 66–8, 75, 82, 84, 88–9, 98, 111–12, 155, 216
modernity, 6–7, 12, 16, 19, 63–4, 77, 106, 226–7
Mumford, Lewis, 25–32, 105
  aggression, 104–5, 133
  America, 49, 112
  atomic bomb, 99
  automation, 51, 180
  capitalism, 120–1
  Christianity, 66–7
  communications, 142, 155
  environment, 103, 108–9, 217–18
  industrialisation, 78
  influence, 42, 46–7
  organicism, 88, 180, 188–93
  progress, 200
  science, 72
  totalitarianism, 98–9
  see also megamachine
myth, 41, 53, 65, 73

nature, 40, 69–83, 86–8, 101–2, 178, 186–96, 201–5, 214–16, 227
  domination of, 2, 9, 35–6, 50–1, 130–1, 133, 150, 154, 165
  transformation of, 41, 52, 54, 103, 130, 168–9, 171–2, 175–6, 201
  see also human nature, second nature
negative unification, 110, 220–1
neutrality, 6, 9, 15–17, 21n, 54, 97, 108–9, 113, 119, 157, 203, 223, 228–9
New Left, 36–7, 53
nihilism, 37, 39, 43–4, 49, 84, 175, 181, 200–1, 203
nuclear, 38, 52, 82–3, 100–19, 165, 215, 226, 230
  atom bomb, 15, 40–1, 49–50, 83, 85–6, 99–112, 200–1, 218, 226
  fallout, 83, 215, 218

nuclear war, 18, 37, 50, 55, 102, 110, 220–1

phenomenology, 25, 173–4, 181, 222
the planetary, 106, 193, 198–200, 215–17; *see also* Earth, world
pollution, 83, 97, 103, 217
private sphere, 53–4, 124, 144–52, 155–8, 177, 197
production, 65–7, 74, 78, 86, 118–37, 197
  and destruction, 66, 83, 101–2, 106, 204, 213, 221
  industrial production, 76–7, 79–80, 178
  Marx, 44, 46
  media, 148–9, 155–7
  see also automation, labour
progress, 35, 149, 223–4, 230–1
  false progress, 194, 200–1, 205, 214, 224–5
  as ideology, 47, 73, 75, 81–2, 113, 204, 216, 221
  liberal progressivism, 5–6, 10, 15–16, 28, 211–12
  Marx, 33, 44
  optimism, 16, 40, 44, 135
  self-defeating, 79, 84, 112, 125, 158, 117, 228
  technological progress, 4, 29, 50–1, 54, 78, 217–18
propaganda, 53, 143–4, 152–9
public sphere, 53–4, 124, 144–52, 155–9, 177, 197

radio, 45, 131, 143, 149–51, 153, 155–6, 192, 198
rationality, 35–6, 44–6, 75, 87, 158
  Cartesian rationalism, 173
  irrationality, 42, 44–5, 97, 109–10, 133, 227
  logos, 134, 180
  rationalisation, 12–13, 18, 73, 80, 129, 154, 165–6, 203
  technological rationality, 37, 42, 72–4, 79, 105, 109, 134, 143, 199, 205, 223
raw material, 77, 192
  humans, 20, 99, 132–3, 136, 165–6, 172, 180–2, 196, 201
reification, 36, 42, 75, 128–32, 192, 202; *see also* alienation, Lukács, Georg

INDEX

rights, 166–9, 171, 179, 182
romanticism, 32, 42, 49–50, 189
Rome, 39, 63
   technique, 66–7, 178

sacralisation, 33, 67, 170, 176–7
   desacralisation, 33, 165–6, 172, 179,
      182
science, 52, 65–6, 73, 81–2, 154, 163, 171,
      196, 214
   critique of, 35, 109, 175, 177, 181
   relationship with technology, 12–13,
      86–7, 221
   see also scientific revolution
scientific revolution, 68–73, 86
second nature, 129, 191–4, 201, 205, 224,
      227, 231
Second World War, 13, 28, 33, 39–40,
      47–9, 96–7, 104–6, 112
Soviet Union, 5–6, 49, 53, 96, 98, 113,
      178, 200
space travel, 52–4, 193, 200; *see also*
      Earthrise
Spengler, Oswald, 10, 16, 42, 120
standardisation, 45, 123, 132, 143, 151,
      153, 156, 158, 197
Stengers, Isabelle, 211, 226–7, 229

technê, 11, 47, 64–7, 106
technique, 11, 72–5, 107–9, 153–4, 178–9,
      195, 200, 203–4
   in Ellul, 9, 30, 32–3, 51–2, 65–7, 78–9,
      85, 111–12, 145, 199, 203, 217–18
technology
   autonomous technology, 3, 7, 119, 121,
      128–33
   critical idea of technology, 1–20, 27–8,
      50, 55, 96–99, 118–21, 186–7,
      191–205, 211–15, 220–32

machine technology, 3, 120–2
   see also catastrophic technology, technê,
      technique
television, 134, 156, 198
   illusion, 41, 86, 127, 192
   private sphere, 143, 149–50
   propaganda, 153, 155
de Tocqueville, Alexis, 34, 146, 149
tools, 77, 81
   distinct from modern technology, 12,
      119, 121–4, 169, 228–9
totalitarianism
   historical totalitarianism, 16, 49, 96
   and liberalism, 5, 211, 226
   technological totalitarianism, 18, 80–2,
      97–8, 107–14, 123
   totalisation, 9, 113, 145–6, 149, 151,
      156–7, 196–200, 220

utopia, 27, 69, 135
   technological utopianism, 40, 88, 134,
      217

Vietnam War, 113–14, 200

Weber, Max, 7, 10–11, 73
Winner, Langdon, 6–8, 18, 75, 119, 228–9
Wolin, Richard, 37, 43
world, 80–2, 136–7, 147–51, 170–2,
      186–205, 213–16, 220
   pseudo world, 81, 195
   technological world, 8–10, 53, 74–5, 77,
      133, 171, 186–7, 191–6, 203
   worldliness, 9–10, 124–5, 186–205
   see also earth, the planetary, second
      nature

Zimmerman, Michael, 48–9, 65, 67, 77,
      84, 133, 219